Alexander Monro

**The United States and the Dominion of Canada**

Their Future

Alexander Monro

**The United States and the Dominion of Canada**
*Their Future*

ISBN/EAN: 9783337188566

Printed in Europe, USA, Canada, Australia, Japan

Cover: Foto ©Suzi / pixelio.de

More available books at **www.hansebooks.com**

THE

# UNITED STATES

AND THE

# DOMINION OF CANADA:

## THEIR FUTURE.

BY

ALEXANDER MONRO, ESQUIRE,

AUTHOR OF

A TREATISE ON LAND SURVEYING ; HISTORY, GEOGRAPHY, AND
PRODUCTIONS OF NOVA SCOTIA, NEW BRUNSWICK, AND
PRINCE EDWARD ISLAND ; HISTORY, GEOGRAPHY, AND
STATISTICS OF BRITISH NORTH AMERICA, &c. &c. &c.

SAINT JOHN, N. B.
PRINTED BY BARNES AND COMPANY,
PRINCE WILLIAM STREET.
1879,

# CONTENTS.

iv

# INTRODUCTION.

In the following pages we propose to sketch some of the leading topics relating to the future of the United States and the Dominion of Canada. We shall trace the outlines of the principal regions of habitable country in North America, and describe those parts of it which have fallen to the lot of the Dominion, as well as their relative situation, with regard to each other and to the United States. We shall relate the chief resources of these two countries; and also their debt, population and other attributes.

We shall recount some of the errors committed by Great Britain with regard to her North American Colonies. We will show that since the close of the rebellion in the United States—since the social and political re-adjustment of the Republic—the policy of Great Britain with regard to these Colonies has been greatly changed; how that the Dominion of Canada is at full liberty to shape her own future in any way she may please; that it is the destiny of small and comparatively weak and far scattered communities, like the settlements in the Dominion, to be incorporated with Great Powers; that if the fusion of the national communities of England, the union of England and Scotland in 1707, that of Ireland with Great Britain at the beginning of the present century, that of the old thirteen colonies in 1776, that of the two Canadas in 1841, and that of the union of the

Canadian and other Provinces in 1867, were necessary, it is equally necessary that the branches of the Anglo Saxon and other families in North America, should unite with each other, as predicted by the Right Honorable John Bright: "For one vast Confederation stretching from the frozen north in unbroken line to the glowing south, and from the wild billows of the Atlantic westward to the calm waters of the Pacific main. Such a confederation," said Mr. Bright, "would afford at least some hope that man is not forsaken of heaven, and that the future of our race may be better than in the past."

We shall show that the United States and Dominion of Canada belong as it were to each other—that they are the geographical and commercial complement of each other; that their union has long been an open question, which has been frequently discussed in the Provinces and States, and by many leading statesmen in Great Britain. How that the ties and interests which bind these American nationalities to each other are much stronger than those which link these Provinces to the British Empire. And how that the Provinces are utterly defenceless in the event of war with the United States.

To those who have read the former works of the writer relating to these Provinces, and who may peruse the following pages, a word of apology is due.

Having carefully noted the development of the resources of the country, and also the official and other reports representing vast areas of unoccupied lands, suitable for settlement in nearly all the Provinces and Territories of British North America, the writer was led to believe that this country was capable of being formed into a nation—a nation able before now to sustain national obligations; and was highly pleased to entertain this view of the future. Having, however, collected the most reliable information in relation to the geography and resources of this country, and, having, as a land surveyor and otherwise had occasion, during the last forty years, to traverse large sections of some of the Provinces, we have learned that but little reliance can be placed on many of the statements made with regard to the resources of British North America. It is now obvious that this immense region, with the exception of a few isolated and comparatively limited areas, is not fit for settlement. And, therefore, acting under the dictates of conviction, we feel it a duty to abandon ground which is no longer tenable.

Apart, however, from these views of the future, the work, we hope, will be found to contain much useful information relating to the United States and British North America. We have placed together an account of the chief resources of both countries, and thus enable the reader to compare and draw conclusions which cannot fail ere long to engage the serious attention of their inhabitants.

A. M.

Port Elgin, New Brunswick,
February, 1879.

## rth America United-G... and
### Ireland ...Federal.

'O THE EDITOR OF THE ...

ms to be directing the forces which are work-
; the di... Integration and dissolution of the
nadian confederation and the ...fication of
continent under one flag ... one central
thority in such a manner as to insure its early
summation. Since I became an earnest stu-
nt of Canadian politics, in 1864, the means
d methods adopted by the party which has
ated, directed, and controlled legislation and
administration of the law, with the excep-
s of one term of Parliament (from 1873 to
78, to maintain, as was asserted, British con-
tion, have exerted a contrariwise effect.

The first fatal mistake made by the Crown,
ough the person of Sir John A. Macdonald,
s its insistence in the retention by it (the
wn) of the right and power to veto legisla-
n enacted by the local Parliaments of the
vinces. It left an open door for legal but
ustifiable assaults upon provincial rights.

n 1844 Canadians were nearly all free traders
i the American tariff was a stumbling block
he way of political union with us. Sir John
cdonald adopted a protective tariff and at
t removed our tariff as an objection to con-
ntal union. He built a transcontinental
way from Halifax to Vancouver at the cost
he State, and in so doing created a debt 400
ent. greater per capita than the debt of the
ed States, and increased the per capita cost
overnment and defence nearly $8.00, against
than $4.00 in this country, if we deduct the
t of the army, navy, foreign intercourse, and
sions, for which Canada does not provide.

e abolition of the State church, the secret
ot, one day's polling, the holding of elections
he same day in all ridings, representation
ording to population, the creation of town-
, town, city, and county councils, the re-
tion of the financial qualification for the ex-
se of the franchise to a nominal amount,
ersal education at the cost of and under
direction and control of the State, the
ption of the Federal principle in creating
Dominion, a decimal currency, the net ton
he legal ton, and last, but not least, the
tion by the courts of Canada of legal de-
ns given or confirmed by the higher courts

...mbined with an ...
p this unhappy, unwise
f to ... garter, a d
ard, also our nor...
...in is pushing and ...
...dized. Outside of the ...
not do more where under the Brit
army Canada by calling her nav..., ...
of t..e same how she ...
devils d will. How bitter ...
the older children and the mot...
interests of Newf.und'.nd are with ...
lic, and after her rearrage with Can...
been consummated she will realize ...
than she does now and will see ...
will make it unpleasant for the c...
if it is d nied to her. The... an in. ...
gotiate with the Dominion for ... ti
with this rep if lic it will me... th. ...
America. We shall not be co... ...
two marriage contracts. This is no ...
ter, as we can make one job of t...
Jonathan will have to marry only ...
therefore cannot be charged with mak...
amous marriage, as Jacob did in ...

It is very kind of Mother England ...
and improve her fortifications ... ...
Quebec, Bermuda, Kingston, and ... (
and construct others at Esquimault and
nect them all by wire and cables. In d...
they will all come under the Americ...
surely as the sun rises in the east and -
the West and rain descends upon the j...
the unjust. The forts at San Lucia ...
ston, Jamaica, will protect the entran... ...
Gulf of Mexico and Nicaragua Canal ...
other canals as we may construct th...
tral American States. It will be inter...
watch the Newfoundland baby when ...
home in Ottawa. She has not control ...
the sea and tides for three hundred yea...
nothing; she knows her rights and dar...
tain them or smash the colored glass ...
windows of the Capitol. When she gets ...
at Ottawa for the session it will be ...
than tobogganing to watch her. I w...
farce played by British and Canadian
knights with pleasure and satisfaction, ...
I see as far as Jonathan is concerned ...
game of "heads I win and tails y... l..se." T
fore he can see her dowry increas d ...
fortifications enlarged and well supplied
perfect equanimity.

Under the Washington treaty we are d...
our claim for consequential damages. ...
Sumner once proposed that we should take
ish North America in settlement for them,
it looks at the present time as if his sugge
might be adopted.

The most expeditious way for the con...
tional Irish party in Great Britain to se
home rule for Ireland is to permit the Tory
to obtain power, and for the friends of Ir
here who support the anti-dynam ite Irish
in Great Britain to exert their influence in c
possible constitutional manner to secure t
litical union of the United States and Car
The end of the monarchy would soon follow
Great Britain would become a federal repu
with Ireland, Scotland, England, and Wal
sovereign States in a federal union. Ira
would then secure much more than she no...

# THE FERTILE AND INFERTILE REGIONS OF NORTH AMERICA.

UNDER this head we purpose, in a brief article, to trace the outlines of the fertile and infertile regions of North America; and particularly, those relating to the United States and the Dominion of Canada.

The resources of North America are so vast, and the spirit of progress so broad, that it is not easy, at this early stage of its history, to mark out the future with any degree of precision. Every year agricultural operations are being extended over greater areas of highly fertile lands, and new mines of wealth are being opened; hence commerce is being extended far and wide. There are many circumstances and events which the hand of time carries along that tend to disarrange our calculations. It is only by estimating the probabilities as well as we can, after we have measured the resources, and marked the progress made, and by taking the balance of probability as a guide, that we may arrive at a reasonable approximation.

There is no part of the world where the surface of the country, and the conditions of climate and production, are so varied as in the habitable part of North America. The United States in the centre, is bounded by the Dominion of Canada on the North, and by Mexico on the south; and each is bounded on the west by the Pacific Ocean, and on the east by the Atlantic. The lands adjoining the Pacific, as well as those bounding on the Atlantic, the high table lands of Mexico, the low lands of its Gulf coast, the great plains of the west, the States and Provinces in the east, and the Arctic Slope of the Dominion, have each conditions of climate and production which differ very much from either of the others. In a word, the drought of the west and humid atmosphere of the east, the heat of the south and the cold of the north, indicate great extremes. The effect of these extremes on the human system are equally marked. The fevers prevalent in the Gulf States, and the pulmonary complaints common in the Eastern States and Provinces, are not known in the State of Colorado and other high lands of the West. Indeed, the colored race, in their cotton plantations—adjoining the

Gulf of Mexico—may be happy in the midst of fevers, and under the burning heat of the southern sun. The Icelanders might feel equally at home and equally happy amidst the horrors of a Winnipeg winter; and the Esquimaux prefer remaining in the icy regions of the North. But these extremes are not congenial to the Anglo-Saxon and other progressive races of the human family, who prefer the more happy medium which this great region possesses. Hence the more central regions between the Arctic Slope and the Gulf of Mexico, and spreading eastward towards the Atlantic Ocean, and southward over Mexico, will be the home of the great family of mankind in North America.

And the differences in climate also indicate difference in the products of the soil. Cotton, rice, sugar and tobacco are peculiar to the South. Indian corn cannot be raised to advantage as far north as the Dominion of Canada; and wheat, except in the Ontario peninsula, is but a small crop in the Eastern States and Canadian Dominion.

The destiny of North America may be said to be determined on the west by the "Great Desert," and on the north by the Laurentide Mountains. These boundaries, which are well defined, divide the fertile lands from the great infertile regions of the west and north.

As the mountain ranges influence the climate, and limit in a great measure the extent of fertile lands, a brief description of them cannot fail to be interesting.

The United States and the Dominion of Canada are traversed in a north and south direction by three ranges of mountains, besides subsidiary ranges; and crosswise, between the Gulf of St. Lawrence and the centre of the continent, by the Laurentide Mountains just referred to.

The eastern range, generally known as the Appalachian or Alleghany Mountains, rise in the Gaspé Peninsula, south of the St. Lawrence, in the Province of Quebec, and extend in a system of parallel ridges, in a south-westerly direction, and nearly parallel to the Atlantic Ocean, about 1,300 miles; passing through the States of Vermont, New York, Pennsylvania, Maryland, Virginia, the two Carolinas and Tennessee, into Alabama. The summits of this range are from one thousand to four thousand feet above the sea. The range has various local names: Near the City of Quebec, it is known as the Notre Dame Mountains; and farther south, as the New

Hampshire Mountains; in New York, as the High Lands: in Pennsylvania, as the South Mountains; and in the other States, generally as the Blue Ridge. In Canada, the Gulf and River St. Lawrence separate the range from the Laurentide Mountains. And south-westward, the Hudson, Delaware, Susquehanna, and other Atlantic rivers pass through the chain. It was on the eastern slope of this range, from the Gulf of St. Lawrence southward, where the chief part of the .early settlements, north of Mexico, were formed. It is now inhabited by a large population. And, though large areas of the country are unfit for cultivation, yet there are numerous and extensive farm districts yielding largely. Many populous cities and towns, numerous seats of manufacturing industry, and vast accumulations of wealth, characterize this section of the American continent.

On the opposite side of the continent, near the Pacific ocean, there are two ranges of Mountains, known in California as the Coast and Sierra Nevada ranges. The former has its southern terminus at the bay of San Francisco, and follows the Pacific coast northward through the States of California and Oregon into Washington Territory, where it unites with the Sierra Nevada range. In California the latter is about one hundred miles east of the coast, and observes a course nearly parallel to it. It is about seventy miles in width, and is known in Oregon, Washington Territory, and in British Columbia, as the Cascade Mountains. The coast range in California is divided into a number of nearly parallel ridges, with an aggregate breadth of about forty miles. Between these ridges there are extensive valleys of highly fertile lands. Between the coast and Sierra Nevada ranges lie the Sacramento and San Joaquim valleys, extending north and south, one valley being a virtual continuation of the other. It is known as the Great Valley, and is about 450 miles in length, by an average width of about forty miles. These valleys are traversed by numerous streams, which afford sufficient water to irrigate large sections of them in the event of droughts. And the Sierra Nevadas comprise two parallel ranges, with numerous valleys between. All the valleys near this part of the Pacific coast converge as they extend northward, and ultimately disappear in a sea of mountains. In California, these valleys contain about forty millions of acres of the most fertile lands to be found on this continent.

The summits of the coast ranges are about eight thousand feet, and some of those of the Nevada and Cascade Mountains, from ten to fifteen thousand feet above the sea.

The dominant range of the continent is the Rocky Mountains. It extends from the Gulf of Mexico northward far into the arctic regions, where it unites with all the other ranges on the Pacific side of the continent. The base of this range varies in width from twenty to sixty miles, and its summits are from eight thousand to sixteen thousand feet above the sea. On the line of the Central Pacific Railroad, this range is about nine hundred miles east of the Sierra Nevada Mountains; in British Columbia, these two ranges are only from one hundred to one hundred and twenty miles apart. Sanford Fleming, the chief engineer of the Canada Pacific Railroad survey, says: "The Rocky Mountain zone observes a general parallelism with the Pacific coast, and in British Columbia is from 300 to 400 miles distant from it. These mountains rise like a colossal wall above the continental plain on its eastern side." In British Columbia the coast and cascade ranges being one, "extends along the entire sea-board." It "rises abruptly from the sea level, presenting from the water an extremely bold and defiant aspect." The ocean front of the mainland of British Columbia is penetrated from twenty to sixty miles inland by a great number of excellent harbors. The breadth of the Cascade Mountains from the head of the harbors varies from one hundred to one hundred and twenty miles, and they rise to a height of from five thousand to eight thousand feet above the sea.

Between the Rocky and Cascade Mountains there is a vast inter-alpine plateau of a wedge-like form, narrowing as it extends northward. Its height in British Columbia varies from 3,000 to 4,500 feet above the sea. This plateau is traversed by minor ranges, known as the Cariboo, Selkirk and Gold Mountains. And Vancouver Island is traversed by mountains in all directions. Indeed, British Columbia is designated as a "Sea of Mountains."

The height of the plateau in the United States is also great. The Salt Lake basin, which is a depression in the great plain, is about 4,300 feet above the sea. This anomalous basin is said to be 359 miles in length by an average breadth of 180 miles. It contains the Salt Lake and the Mormon Settlements.

But the most important range of mountainous country in North America is the Laurentide region. Important, because it limits the area of fertile lands to the northward, and in an east and west direction for a distance of about two thousand miles. Beginning at the Labrador coast, it extends along the north bank of the Gulf and up the River St. Lawrence to within twenty miles of the City of Quebec: it sweeps to the northward of this city, and the City of Montreal, where it is about thirty miles north of the St. Lawrence. "Beyond this," says Sir W. E. Logan, "it extends up the Ottawa on the north side, for about a hundred miles, and sweeps round thence to the Thousand Islands, near Kingston; from which it gains the southern extremity of Georgian Bay, and continues along the eastern and northern shores of Lake Huron and Lake Superior." It continues westerly to the Red River valley, a little west of the centre of the continent, and thence turns northward around the east shore of Lake Winnipeg, where the range is comparatively low. It terminates in the arctic regions, a distance of 3,500 miles from Labrador.

The Laurentides, says Sir William, "occupy by far the larger portion of Canada," that is, the Provinces of Quebec and Ontario. The range in the Ottawa region is about two hundred miles in width. "The area occupied by the Laurentian series in Canada," he says, "is supposed to be about 200,000 square miles."

With the exception of an off-shoot of this range, which crosses into the State of New York, known there as the Adirondack Mountains, which cover an area of six millions of acres in that State, and some out-lying spurs at the west of Lake Superior, the Laurentian formation is in the Dominion of Canada. This arm of the Laurentides leading to the Adirondack Mountains divides the Canadian side of the valley of the St. Lawrence into two nearly equal parts. That between the City of Ottawa and the United States boundary, extending to fifty miles, is a triangular plain, "comprising," says Sir William, "about 10,000 square miles, being nearly level, and of a good agricultural character." And, though the surface of the western peninsula of Ontario, or "upper plain, occupying about 10,000 square miles, has a generally smooth surface, it swells into a height which is not inferior in elevation to some of the highest points in the more rugged Laurentian country between Lake Huron and the Ottawa." Though for convenience, we designate this formation as Laurentian, they are known in some places as the Huronian series of rocks. The Laurentides are said to be the oldest series of chrystalline rocks in the world. The summits of this region vary in height from 4,000 feet below the City of Quebec, to 1,500, 1,700, and to 2,300 feet above the sea, in the Ottawa and Lake Huron District. In the Lake Superior region they reach a height of about 2,000 feet above sea level. The Saguenay, Ottawa and other tributaries of the River St. Lawrence, penetrate the range for considerable distances. This mountain region between the Gulf of St. Lawrence and the Red River Valley, forms the water-shed between the St. Lawrence and Hudson's Bay.

Within the folds of this elevated plateau there is said to be more than a thousand lakes, some of them of considerable size. Lake Nipissing covers a surface of 294 square miles, and is 639 feet above the sea; Grand Lac has a surface of 560 square miles, and is 700 feet above the sea; and the Lake of the Woods is 1,042 feet above sea level. There are other lakes in this region containing equally large areas, and extensive swamps are numerous. The River valleys in the Laurentian region are generally narrow. The climate, in consequence of the great elevation and northern aspect of the chief part of the country, is unfavorable for agricultural operations, even if the land was suitable. Indeed, the great Laurentian region, except a few isolated spots, will ever remain outside the pale of a habitable and food producing country. Hence, the southerly flank of this region in Canada, between the Gulf of St. Lawrence and the Red River valley of the north-west, a distance of two thousand miles, is the northern boundary of food-producing North America.

The unfortunate position of this region of rocks so far south, or in other words, the international boundary being so far north, can hardly be realized at present, especially in regard to the future of the Dominion of Canada. This will more fully appear as we proceed.

Having thus briefly traced the chief mountain lineaments of this part of the continent, we shall, in the next place, trace the outlines of the great fertile region of North America.

With the exception of two narrow belts of highly fertile country, one adjoining the Atlantic, the other the Pacific,

Mexico consists of lofty tropical plateaus, which rise in step-like forms into broad terraces, sloping upwards, each presenting the climate and productions of the temperate zone. The highest level is a mountain chain of great elevation.

A road from the City of Mexico, the ancient capital of the Montezumas, northward for a thousand miles, would run over a country hardly varying from an altitude of 7,500 feet above the ocean. Mexico contains a large proportion of highly fertile land. Its mineral and other resources are also very great.

But from the day on which Spain set her foot on the soil of Mexico to the present time, the benevolent dispositions of Providence for the happiness of his creatures have been counteracted, first by Spain, and since by her offspring. The history of Spain in Mexico is a terrible one. She has stamped her worthless civilization on one of the most generous soils and genial climates in the world. Since 1810, when Mexico no longer able to bear the burdens imposed by her task-masters, raised the standard of revolt, anarchy and bloodshed has been Mexico's annual contribution to the pages of history. Indeed, she has only exchanged the mis-government of Spain for hopeless anarchy. In place of having twenty millions of educated and progressive people, the population of Mexico is only about ten millions, the greater part of whom are but half civilized. Mexico and Central America are well adapted to sustain a large population by the products of the soil.

The habitable and generally fertile region of North America, including the regions referred to, is advantageously situated. On its front it has the Atlantic Ocean as far north as the Gulf of St. Lawrence. On the north it is bounded by the Laurentide Mountains before described, and on the west by the American desert. The extreme westerly boundary of the fertile lands of the continent, beginning at the Gulf of Mexico, and proceeding northward, follows a line varying in its inflections between the ninety eight and one hundredth meridian as far north as Lake Winnipeg, in the Canadian north-west. The country thus described is equal to about half the area of the United States, or about equal to one-fourth the area of that part of the continent north of Mexico. Near its northern boundary, the Gulf, River and Lakes of the St. Lawrence afford a passage for vessels to the heart of

8

the continent. From the south this region is penetrated by the Mississippi river and its extensive and wide-spread branches, which in aggregate length afford about 20,000 miles of navigation.

Though the region east of the one hundredth degree of longitude embraces nearly all the fertile lands climatically adapted for settlement north of Mexico, it also includes large areas of unproductive lands. The Appalachian mountains cover a large area; and the Eastern States and Provinces contains a large proportion of worthless lands. The Gaspé Peninsula, the Adirondack Mountains in the State of New York, and many smaller tracts of infertile lands, deface the country.

But in contradistinction to the country north of the river and northern lakes of the St. Lawrence, and the region between the one hundredth degree of longitude and the Pacific Ocean, there are vast areas of land of the highest fertility. The valley of the Mississippi alone is estimated as capable of producing food sufficient to sustain the population of Europe. According to the Report of the Commissioner of the General Land Office of the United States for 1876, the country " between the eastern boundary of the State of Ohio, and the central portions of the States of Kansas and Nebraska, covering the valleys of the Ohio, the Mississippi and the Missouri rivers, and extending from the eighty-first degree to the ninety-fifth degree of west longitude, is a region well classified as the fertile belt of the continent." As near as we can ascertain from the reports and maps before us, this belt has a breadth of thirteen degrees of longitude; and a length between the Gulf of Mexico and the Lake Superior region of about 1000 miles, and contains upwards of four hundred and fifty millions of acres. Fifteen entire States, namely, Michigan, Ohio, Indiana, Illinois, Wisconsin, Kentucky, Tennessee, Mississippi, Alabama, Georgia, Minnesota, Iowa, Missouri, Arkansas and Louisiana, besides parts of other States, lie within its boundaries. Ten of the fertile States lie on the east side of the Mississippi River.

If to this region of fertile land we add that in the valley of the River St. Lawrence, and the fertile lands south of the Alleghany Mountains, including the Atlantic Provinces of the Dominion of Canada, the area of productive land in North America is truly immense. Besides, in the old States and Provinces there are large areas of second-class soils, which are being filled in from adjacent settlements.

In the foregoing enumeration of fertile and productive areas, we have not included the Canadian region lying west of Red River, and north of the great American Desert. The Report before referred to, says: "In all that section lying between the one hundredth meridian on the east, and the Cascade Range and the Sierra Nevada Mountains on the west, and, within these limits, from the Mexican line on the south to the International boundary on the north;" and we add, in all the region north of this boundary, "a totally different set of conditions, geographical, physical and climatic, is found to exist. Within this vast area agriculture, as understood and pursued in the valley of the Mississippi and to the westward, has no existence." Between the valley of the Mississippi and the Pacific, "irrigation is indispensable to production. That there are limited areas within which by its aid crops are and may be secured is true, but the proportion of land within the area now treated of, which under the present system of disposals, can by this means be made productive, is insignificant. Under a system which would justify large expenditures and ensure the utilization for the purposes of irrigation of the whole volume of water reaching the valleys from the mountain streams, but a mere fraction of the whole great area could be made fit for tillage." This infertile region has a breadth of 700 miles on the international boundary, lat. 49° north, and extends northward into the Dominion of Canada, a distance of from 200 to 250 miles.

As the Canadian plains situated north of the desert possess a different set of conditions from those of the Mississippi valley, we purpose, further on, to devote more space to their consideration.

The great mountain ranges on the Pacific side of the continent arrest the water-clouds from the west on their way to the interior, and thus prevent precipitation of atmospheric moisture on the plains. In the southern part of the great Californian valley, the rain-fall is so light that two crops in five years is all that may be expected. The yield in fruitful seasons, however, is immense. Some seasons the northern parts of the valleys suffer for want of sufficient rain-fall. Of the great valley, 7,650,000 acres are susceptible of easy irrigation. In all, 12,000,000 acres might be irrigated, and thus ensure,

2

in all seasons, an immense yield of crops; as the lands are
among the best on the continent.

Unlike the Atlantic coast there are but few Rivers on
the Pacific side; consequently there is only here and there
and far between, a narrow valley fronting the seaboard where
agricultural operations can be pursued with profit. A very
large part of the country north of California is a consolidated
sea of mountains. The most important agricultural region
north of California is in the valley of the Columbia River.
This river, for a considerable distance from its mouth upwards, forms the boundary between the State of Oregon and
Washington Territory. The wheat product of the Columbia
Valley is now considerable. The country is being rapidly
settled. The climate of this part of the coast is excellent.
The precipitation of atmospheric moisture accumulates rapidly
along the Pacific coast from south to north. At the north end
of Vancouver Island the rain-fall is said to be very great. The
climate as far north as the north end of this Island is highly
favorable to agricultural production; the winter is short and not
severe, while summer is equable, and the whole season conducive to health. But the great want on this coast is sufficient
land fit for cultivation.

The mineral resources are highly valuable. Indeed the
whole region fronting on the coast is a gold field. The mountain slopes and river sands are known to contain large quantities of gold, which as time passes will be more fully developed.
Coal is abundant on Vancouver Island, and in Washington
Territory, as well as iron ore and other minerals useful in
commerce. The mountain sides are clothed with useful forest
wood; and the rivers and sea-board teem with valuable fish.

· The following table, showing the progress of population,
affords an index to the sections where the largest areas of
fertile lands are to be found:—

| FERTILE AREAS. | POPULATION IN | | |
|---|---|---|---|
| | 1850. | 1860. | 1870. |
| California ............... | 92,597 | 379,994 | 560,247 |
| Oregon .................. | 13,294 | 52,465 | 90,923 |
| Washington Territory...... | .......... | 11,594 | 23,955 |
| British Columbia.......... | .......... | .......... | 10,586 |

The great plateau adjoining the Sierra Nevada and Cascade ranges on the coast is about 4,000 feet above the sea. This desert, as already shown, is unfit for cultivation, except in low river valleys; hence, it is only near the ocean where there are any tracts fit for settlement. The Utah Basin is an exception. Notwithstanding the aridity of the soil, and the oft-repeated visits of the locusts, the population of Utah has increased from 11,380 in 1850; 40,273 in 1860; to 86,786 in 1870.

It may finally be assumed, that with the exception of a few small and isolated spots, the country between the one hundredth degree of west longitude and the Pacific Ocean is unfit for cultivation.

The great plains of the Far West, however, are not altogether without value to the agriculturist. Large areas of the western region afford, without the aid of man, a highly nutritious pasturage, where herds of sheep, cattle, and horses can be kept for the greater part of the year. The official report, from which we again quote, says:—"The excellence of the pasturage of the plains and valleys consists in the fact that the grasses, though thin and of slow growth, retain their nutritious qualities throughout the entire year, and in the further fact that, for the present, the range is only limited by the possibility of reaching suitable watering places."

With this brief description of the country south of the Laurentide Mountains of Canada, and west to the Pacific Ocean, we purpose to view the country on each side of the international boundary, comparing by the way, the share of habitable country which has fallen to the lot of each of these nationalities; comparing also their progress in population, and in the development of their resources.

# THE UNITED STATES.

The greatest width of the United States from ocean to ocean is 3,200 miles; and 2,000 at its narrowest part. The average breadth is 2,500 miles. From the Canadian boundary southward to its southern extremity, the distance varies from 1,200 —1,300, to 1,600 miles. The ocean frontage of the United States extends along the Atlantic, a distance of 1,700 miles: 1,400 on the Gulf of Mexico; and 1,260 miles on the Pacific ocean, making a total of 4,360 miles, exclusive of indentations. This vast extent of sea-board is navigable at all seasons.

The Colorado River, which discharges into the Gulf of California in Mexico, has a length of one thousand miles in the United States. The Mississippi River takes its rise near the Canadian boundary, the parallel of 49° north latitude. From the head of the Missouri, emphatically the main river, though designated a tributary, it traverses a distance of 4,491 miles.

The Ohio enters the Mississippi at a distance of 1,145 miles from the Gulf of Mexico; it is navigable for a distance of 975 miles, and during periods of high water, vessels ascend 200 miles farther. The Mississippi with its wide-spread affluents drains an area of 785,000,000 acres, or nearly one-half the area of the United States, and discharges into the Gulf of Mexico. It traverses ten States.

The Union is divided into thirty-eight States and ten Territories, exclusive of Alaska. Of these, thirteen front on the Atlantic Ocean; five on the Gulf of Mexico; three on the Pacific Ocean; and thirteen bound on the Dominion of Canada. Of the latter, nine are east of the great desert, or within the boundaries of the productive regions, and comprise about 250,000,000 of acres, and contain a population of about seventeen millions. That part of the United States situated between the Atlantic and the desert is estimated to contain about seven hundred millions of acres; and is divided into thirty States; it includes, besides, a part of those adjoining on the west. The fertile lands in the Union drained by the river and lakes of the St. Lawrence are comparatively limited. It drains about forty-four millions of acres in the States, a part of which is mountainous. The greater part, however, is highly fertile and densely populated. The table lands lying between the valley of the St. Lawrence and that of the Mississippi, are but slightly elevated above the general level of the country.

In Canada, the St. Lawrence drains an area estimated at two hundred and eleven millions of acres. This immense region with the exception of a narrow strip fronting on the St. Lawrence, is within the Laurentide mountain range.

Thus, we have three slopes or valleys lying within the fertile region in this part of the continent; the Mississippi, St. Lawrence and Appalachian. Nowhere on the face of the globe is there such a vast expanse of fertile lands as these valleys contain. With the exception of a comparatively limited area of mountainous country, it is one vast continental plain, extending from the Atlantic Ocean westward, to the one hundredth degree of west longitude, and in a north and south direction from the Laurentide mountains to Mexico. It is now inhabited by about fifty millions of people, including four millions of Canadians; still there is room for hundreds of millions. Referring to that part of it, in the west, lying south of the Canadian boundary, and which might conveniently adopt navigation by the St. Lawrence as an outlet for the surplus products of the west, John Page, Chief Engineer for the Canadian Board of Works, in his Report for 1874, says:—"The line of settlement is yet a long way from the western boundary of the fertile region, and it is stated on good authority that even in that part of it which furnishes the supply, there are not more than one-fifth of the available lands under cultivation." This food-producing region embraces every variety of climate, from the genial south to the winter cold of Quebec and northern Minnesota. Here the different nationalities of the world may find both climate and soil to suit them. And when Mexico takes her place among the more civilized and progressive communities of this continent, the sea of human life may be indefinitely extended.

But the great barrier to western progress is the desert.

Professor Henry, a very distinguished man in the United States, says:—

"The whole extent of country to the west, between the ninety-eighth meridian and the Rocky Mountains, called the great 'American Plains,' is an arid desert, over which the eye may wander to the horizon without seeing anything to relieve its monotony. . . . And perhaps we shall surprise the reader by drawing his attention to the fact that this line, which is drawn southward from Lake Winnipeg to the Gulf of Mexico, divides the surface of the United States into two very nearly equal parts. When properly understood, this statement will serve to dissipate some of the dreams, regarded as realities, about the destiny of the western part of the continent of North America; but truth in the end takes precedence of praiseworthy patriotic sentiment."

The treeless desert is estimated by G. M. Dawson, F. G. S., to contain six hundred thousand square miles, about one-third of which is in the Canadian Northwest. It extends northerly into Canada, a distance of about two hundred and fifty miles, and lies between the fertile lands on the North Saskatchewan and the International line.

In his "Sketch of the North West America," in 1871, Bishop Tache, who resided in the Canadian Northwest for nearly a quarter of a century, says, page 10:—

"Here is a desert—an immense desert. It is certainly not everywhere a plain of moving sand, and quite dried up; but it is quite vain to think of forming considerable settlements on it. Prairie hay (*systeria dyetaloides*) is almost the only plant which is seen growing on its arid soil. A narrow border of alluvial soil marks its water courses, and these are dry nearly throughout the year. The prairie hay supplies pasturage of the best kind; not only the buffalo delights in it, but horses and other draught animals are very fond of it. This herb, barely six inches high, of which the plants grow so sparsely as to leave the sand or gravel on which it grows everywhere visible, preserves its flavour and nourishing power, even in the midst of the rigors of winter, to such an extent that a few days grazing on one of these remarkable pasturages suffices to restore horses worn out by work to good condition.

"Beyond this advantage, and the game to be found there, I do not know of anything on this vast plain which could attract the attention of economists. The wearied eye seeks in vain for a shore to this ocean of short hay. The weakened traveller sighs in vain for a stream or a spring at which to quench his thirst. The heavens, dry as the earth, hardly ever grant their dews and beneficent showers. The dryness of the atmosphere aids the aridity of the soil; some places of which the geological formation would appear to favor vegetation,

produce no more than the naturally sterile ground. One
travels across this desert for days and weeks without seeing
the smallest shrub."

. It is generally admitted that the destruction of forests tends
to decrease the rain-fall, and render the country less productive in an agricultural point of view. This is now known to
be the case in parts of Europe and America. In California
during very dry seasons, artificial means of irrigation have
been adopted with great effect. Dr. Hayden, in his Geological
Report, in 1867, says, with regard to parts of the American
desert, that "the settlement of the country, and the increase
of the timber, have already changed for the better, the climate
of that portion of Nebraska lying along the Missouri, so that
within the last twelve or fourteen years, the rain has gradually
increased in quantity, and is much more equally distributed
throughout the year." And in 1870, he said: "It is true that
over a width of one hundred miles or more, along the Missouri
River, where the little groves of timber are extending
their area, springs of water are continually issuing from the
ground where none were ever known before; and that the
distribution of rain throughout the year is more equable."

The process of planting forests is a slow one; consequently,
many generations will pass away before any great area of the
treeless plains of the west can be made to sustain any considerable population from the products of the soil. However,
many parts of the plains afford a rich pasturage, and where
sufficient water can be obtained, thousands of cattle, horses,
and sheep, may find plenty of highly nutritious grass.

But the infertility of the western plains is not the worst
feature of the country. The great plague of the West and
Northwest is the locusts or grasshoppers. From the high
plains, near the eastern slope of the Rocky Mountains, they
wing their way in countless numbers over great areas of the
fertile regions to the eastward, destroying almost all kinds of
vegetation in their course. They seem to have been more
destructive in recent years on both sides of the international
line. No satisfactory means have yet been adopted for their
destruction. The winters of the arctic slope seem to be no
obstacle to their reproduction. Nearly every section of the
great prairie region, adjoining the sand plains, has suffered at
times by this plague. In Utah Territory and some other
places, where, in consequence of their ravages, the ground has

been replanted, they have been known to cut the plants down for the fourth or fifth time in one season. Hence, the future of the Red River and Saskatchewan valleys, where the summer season is too short, when one crop is destroyed, to allow a second crop to be matured in the same season, is not hopeful, if the locusts continue their ravages.

Table shewing the population of the United States and British North America for the years named therein:—

| UNITED STATES. | | BRITISH N. AMERICA. | |
|---|---|---|---|
| YEAR. | POPULATION. | YEAR. | POPULATION. |
| 1790.... | 3,926,214 | 1791.... | 279,000 |
| 1800.... | 5,308,483 | 1801.... | 342,000 |
| 1810.... | 7,239,881 | 1811.... | 479,000 |
| 1820.... | 9,633,822 | 1821.... | 790,000 |
| 1830.... | 12,866,020 | 1831.... | 1,200,000 |
| 1840.... | 17,069,453 | 1841.... | 1,656,700 |
| 1850.... | 23,191,876 | 1851.... | 2,487,855 |
| 1860.... | 31,443,321 | 1861.... | 3,294,654 |
| 1870.... | 38,558,371 | 1871...... | 3,730,774 |

The population of the United States may not exceed 47,000,000 in 1880; and that of the Dominion, 4,100,000 in 1881.

The population in the last table is exclusive of Indians; of whom, in 1877, there were 250,809, in the United States; and 102,000 in the Dominion of Canada in 1871.

The amount expended by the United States in 1877, for education of the civilized tribes, was $337,379; of which $209,337 was paid by the Central Government. The number who learned to read during the year was 1,206; and the total number who can read is 40,397.

During the year 1877, some thousands of Indians removed from the United States to the Canadian Northwest. Thus the Indian population of the Dominion is comparatively large.

Previous to the present century, the progress of North America was very slow. The long time that elapsed between the occupation of Quebec by France, the landing of the Puritans, and the conquest of Canada by Great Britain, and the year 1776, when the United States declared its independence, was a comparative blank in the scale of progress. Doctor Franklin estimated the population of the "Old Colonies" at 1,200,000 in 1775; and the remaining Provinces of Britain contained about 80,000 at that date.

The aggregate area of the old thirteen colonies is 218,829,600 acres; and their population in 1870, was 16,433,672. The total area of the United States, including Alaska, purchased from Russia, is 2,291,352,320 acres. In his report for 1877, the Commissioner of the General Land Office, estimates the population of the Union "at 46,000,000." Considering the decrease in immigration, probably forty-five millions would be the most correct estimate.

The Anglo-Saxon population was but small in Europe previous to the present century. The population of Scotland in 1707, the date of her union with England, was less than one million; while that of England and Wales was only about six millions. And the population of Ireland in 1800, the date of her union with Great Britain, was about five millions; while that of the latter was ten and a half millions. Hence, the total population of the British Isles, was, at the beginning of the present century, only fifteen and a half millions.

Though emigration from Great Britain to North America has been very large during the last fifty years, her population in 1871 was more than double what it was at the beginning of the present century, and is increasing rapidly. As she has to depend upon the valley of the Mississippi for a large part of her flour, it is only natural that the redundant population of Britain will continue to remove to the source of supplies,— the valley of the Mississippi, where fertile lands are abundant.

Thus, a country, the chief part of which a century ago was the abode of a large number of pagan tribes, many of them of the lowest type of savage life, has been changed and elevated into a seat of industry, progress, freedom and christianity. There is something sublime in the spread of civilized mankind, with his flocks, arts, commerce, and various customs and industries over the wilderness and prairie regions of this vast continent.

The progress of the country in education, agriculture, commerce, manufactories, ship-building and other pursuits; in the construction of roads, railroads, canals, telegraph lines; in the extension of settlements; in the erection of towns and cities, and in the development of the resources of the country

3

generally, and in the establishment of law and order, has been great indeed, and without a parallel in history.

There is no subject within the compass of these pages which bears more closely on the future than that which relates to the capabilities of the country to sustain human life. It is estimated that North America is capable of sustaining a population of about five hundred millions. The area of fertile land climatically adapted for settlement is immense; and if ever as densely peopled as some parts of Europe and Asia are at present, this is far too low an estimate.

Second only to the agricultural capabilities to sustain human life, are the various products of the deep. Almost every river, lake and sea-coast teem with useful fish. The annual products from these sources are immense.

The commercial resources of North America are only beginning to be developed. Besides the products of the soil and the fisheries, vast areas of the unproductive lands of the country are clothed with forest wood of great value. And many of its mountain chains and hilly regions are richly stored with gold, silver, coal, iron-ore, copper, and other valuable minerals. Indeed, the whole country, from the fur-producing regions of the north to the genial south, and between ocean and ocean, is capable of contributing, in numerous ways, to the support of a vast civilization.

It may be argued, however, from the facts as they exist in some parts of the old nations of the world, that a dense population is not proof that a country is prosperous. In some countries, increase in population means increase in poverty and crime. In the States and Provinces it is otherwise; a large increase in population shows a corresponding increase in production, in wealth, and social progress.

There is, however, a class of persons, chiefly from Europe, who prefer residing in cities, and who, during times of commercial depression, such as that which has prevailed over a large part of North America, as well as over Europe, for the last four years, have suffered for want of remunerative employment rather than reside in the country, where good land is abundant and labor remunerative.

With such exceptions, which are only temporary, there is no part of the world where labor has been better rewarded, population more rapidly augmented, and life more easily sustained, than in these States and Provinces. And in no part of the world does the future present brighter prospects. Indeed, the social future of North America is only beginning to be solved. The comparatively limited areas of cultivable lands in Europe and Asia do not produce sufficient food to meet the actual wants of their rapidly increasing populations. If their inhabitants increase during the next fifty years as in the last, emigration in future must be large. Some will go to Australia; but far the greater part will look to America for homes. And if peace is maintained, and the taxes of the States and Provinces shall be kept on a low scale, immigration to this section of the continent will be large. Statistics of emigration will be found in other pages.

The following table exhibits the progress of surveys and the disposal of public lands in the United States since 1866:—

| FISCAL YEAR ENDING JUNE 30. | NUMBERS OF ACRES. | |
|---|---|---|
| | SURVEYED. | DISPOSED OF. |
| 1867................ | 10,808,314 | 7,041,114 |
| 1868................ | 10,170,656 | 6,665,472 |
| 1869................ | 10,822,812 | 7,666,151 |
| 1870................ | 18,165,278 | 8,095,413 |
| 1871................ | 22,016,607 | 10,765,705 |
| 1872................ | 29,450,939 | 11,864,975 |
| 1873................ | 33,834,178 | 13,030,606 |
| 1874................ | 29,492,110 | 9,530,872 |
| 1875................ | 26,077,531 | 7,070,271 |
| 1876................ | 20,271,506 | 6,524,326 |
| 1877................ | 10,847,082 | 3,440,738 |

The total area of the United States, exclusive of water, and also of Alaska, is 3,002,848 square miles. Of this area, 731,667,583 acres remained unsurveyed up to June 30, 1877. The greater part of this area is unfit for settlement. Less than half the land surveyed during the last ten years is disposed of.

Table showing the quantities of the principal farm products raised in the United States in the years therein named:—

| ARTICLES. | 1850. | 1860. | 1870. | 1876. |
|---|---|---|---|---|
| Wheat, bushels.. | 100,485,944 | 173,104,924 | 287,745,626 | 289,356,500 |
| Indian Corn, " .. | 592,071,104 | 838,792,742 | 760,944,549 | 1,283,827,500 |
| Rye, " .. | 14,188,813 | 21,101,380 | 16,918,795 | 20,374,800 |
| Oats, " .. | 146,584,179 | 172,643,185 | 282,107,157 | 320,884,000 |
| Barley, " .. | 5,167,015 | 15,825,898 | 29,761,305 | 38,710,500 |
| Buckwheat, " .. | 8,956,912 | 17,571,818 | 9,821,721 | 9,668,800 |
| Potatoes, " .. | 104,066,044 | 153,243,893 | 165,047,297 | 124,827,000 |
| Tobacco, pounds... | 199,752,655 | 434,209,461 | 262,735,341 | 381,002,000 |
| Hay, tons......... | 13,838,642 | 19,083,896 | 27,316,048 | 30,867,100 |
| Cotton, bales...... | 2,469,093 | 5,387,052 | 3,011,996 | 4,438,000 |
| Rice, pounds..... | 215,313,497 | 187,167,032 | 73,635,021 | |

The products named in the last table for the years 1850, 1860 and 1870, are quoted from the census reports of those years, and those for 1876 are from the Report of the Commissioner of Agriculture.

According to the last Report, the States which produced the largest yield of wheat and Indian corn were:—

| STATES. | WHEAT, BUSHELS. | INDIAN CORN, BUSH. |
|---|---|---|
| California......... | 30,000,000 | 1,600,000 |
| Illinois............ | 23,440,000 | 223,000,000 |
| New York........ | 9,750,000 | 21,000,000 |
| Missouri......... | 15,240,000 | 102,500,000 |
| Kansas. ......... | 16,510,000 | 82,836,000 |
| Ohio............. | 21,750,000 | 115,000,000 |
| Indiana.......... | 20,000,000 | 99,000,000 |
| Pennsylvania..... | 18,740,000 | 42,250,000 |

Indian Corn is largely produced in all the central and southern States, and in the southern part of Minnesota. The corn crop of 1875 amounted to 1,321,000,000 bushels, being the largest known. The oat crop of that year was above an average, being estimated at 354,000,000 bushels.

"The swamps of South Carolina, both those which are occasioned by the periodical visits of the tides, and those which are caused by the overflowing of the rivers, are admirably adapted to the production of rice....By the introduction of this water-loving cereal, various swamps, which previously had only afforded food to frogs and water-birds, have been changed into the most fruitful fields; so that South Carolina not merely supplied the whole of the United States with all the rice they require, but also annually exported more than a hundred thousand large casks to the various markets of Europe."

The cotton crop is one of the most valuable crops of the United States. It has increased from 2,193,987 bales, at the close of the rebellion, to 4,669,288 bales of 436 pounds each, in 1876. The average yield is about one third of a bale of cotton to the acre. The area of country over which cotton can be produced is very large. There is at least ten of the Southern States of the Union adapted to its production. The largest crop, 4,669,410 bales, was produced in 1859. The entire area in cotton in 1860 was about thirteen millions of acres. An ordinary crop of cotton is worth about two hundred millions of dollars.

The crop in 1877 amounted to 4,485,000 bales. Of this, 2,025,000 bales were exported to Great Britain; 1,025,000 to the Continent; and the remainder, 1,435,000 bales, were consumed in the United States.

# BRITISH NORTH AMERICA.

In the final division of North America, Great Britain has been very unfortunate. It is an undeniable fact that the United States includes nearly all the fertile lands climatically adapted for cultivation north of Mexico. The southerly flank of the Laurentide Mountains is the natural boundary northward of food-producing North America; at least for a distance of two thousand miles—from Labrador to the Red River valley. This being nature's limits, we have no right to complain. But with regard to the International boundary, "No Canadian," says Sanford Fleming, C. E., "can reflect, without pain and humiliation, on the sacrifice of British interests in the settlement that was made." The settlement made by Lord Ashburton, representing Great Britain, and Daniel Webster, the United States Commissioner, "converted," he says, "undoubted British territory into foreign soil," and "alienated the allegiance of thousands of British subjects, without their consent, and made a direct connection on our own soil, between Central Canada and the Atlantic an impossibility." In Europe and Asia British diplomacy has always been managed with a just regard to British interests; but in regard to North America it seems to have been managed at every point in the interests of a foreign power.

"It is evident," says Mr. Fleming, "from an inspection of the map, and from the natural features of the country, that lines of railway might have been projected so as to bring Montreal within 380 miles of St. Andrews, 415 miles of St. John, and 650 miles of Halifax; and that the distance from Quebec to St. Andrews need not have exceeded 250 miles; 67 miles less than to Portland. Fredericton, the seat of Local Government, would have been on the main line to Halifax, and distant from Montreal about 370 miles; and these lines, moreover, would have been wholly within the limits of the Dominion had the international boundary been traced according to the true spirit and intent of the Treaty of 1783.

"The distance between Montreal and Halifax might thus have been lessened nearly 200 miles. St. Andrews would have taken the place of Portland as the winter terminus of the Grand Trunk Railway, and would have commanded, together with St. John, a traffic now cut off from both places, and centered at a foreign port. The direct route would have brought the Springhill coal fields of Nova Scotia some 200 miles nearer Montreal than by the present line of the Intercolonial, and would have rendered it possible to transport coal by rail at a comparatively moderate cost.

"If, under such circumstances, an Intercolonial line to connect the cities of the Maritime Provinces with those of the St. Lawrence had been constructed, the building of 250 miles of railway, representing an expenditure of $10,000,000, would have been unnecessary. Great as this saving would have been, the economy in working it and in maintenance would have been more important. The direct line would also have attracted certain branches of traffic which by the longer route must either be carried at a loss or be repelled. These considerations render the difference in favor of the direct line incalculable, and cause the more regret that the treaty made by Lord Ashburton, which ceded British territory equal in size to two of the smaller States of the Union, rendered such a direct line through British territory forever impossible. Although it is too late to rectify this most fatal error, it is important in a history of the Intercolonial Railway to recount all the steps by which so costly a consequence has been forced upon the Dominion."

New Brunswick lost a breadth of "more than a hundred miles."

Below the city of Quebec the international boundary is, for a distance of sixty miles, only from twenty-six to thirty miles from the river St. Lawrence and the Intercolonial Railroad. Indeed, all the chief cities of New Brunswick, Quebec and Ontario are near the international line. Well might Daniel Webster say that "an object of great importance had been gained to the United States by the settlement of this part of the international boundary."

The Provinces have generally attached blame to the United States authorities with regard to the settlement of this part of the boundary. Mr. Fleming takes a different view: "It is evident," he says, "throughout, that the Executive at

Washington desired to settle the line of boundary described in the Treaty on a fair and equitable basis. Indeed, it is scarcely possible to suggest a proposal more marked by sagacity and justice than that made by President Jackson." And, "had the offers made by the United States been accepted, the boundary would have been satisfactorily established." "The fault," he says, "does not lie with the Washington Government. It is due to the ignorance of the merits of the case, and to an indifference to the interests at stake, on the part of the Imperial representative, who had been entrusted with the protection of the rights and honor of the Empire."

Whatever may have been the views of Great Britain with regard to the future of her remaining possessions on this part of the continent, it is now obvious that in the settlement of her boundary, and other disputes with the United States, the "manifest destiny" doctrine received a powerful impulse. Numerous boundary and other disputes have been handed down as legacies to these Provinces. It is only very recently, by the Treaty of Washington, that some of the disputes have been settled. Between Lake Superior and the Pacific an immense territory, claimed by Great Britain, has been annexed to the United States. And by the award of the Emperor of Germany, the San Juan Islands, and the chief entrance from the Pacific Ocean to the main land of British Columbia, are ceded to the Union.

Consequently, a recent writer very correctly says: "The Dominion of Canada, the legitimate heir to the old French Empire in North America,....has come down to us sadly diminished in extent." However, let us take things as they are, let us trace the boundaries of our habitation, let us ascertain the nature and extent of our resources, and how they can be developed.

A better knowledge of the geographical conditions of the country has enabled us to trace its cultivable parts. We have already said that the southerly flank of the Laurentian mountain region, between the Gulf of St. Lawrence and the Red River valley of the Northwest, a distance of two thousand miles, is the northerly boundary of the cultivable part of this section of North America. By this statement we do not mean to say that there are no places fit for settlement in the Laurentian mountain region. Isolated settlements already

exist, and others may be formed within this mountain plateau. There is a small area of good land near Lake Nipissing
which is being rapidly settled. There are some tracts of
second class soils still unsettled in all the Provinces east of
Red River valley. But the quantity of third and fourth class
soils is comparatively very large in all the old Provinces of
the Dominion. But we say, with A. J. Russell, C. E., in his
work on the "Northwest Territories," in 1869, referring to
the unsettled lands between Lake Superior and the Atlantic:
"That it would be heartless iniquity to induce settlers in
search of permanent homesteads to sink their labour on such
lands when better can be had." As far back as 1862 the Surveyor General of the two Canadas, in his official report, said:
"The whole quantity of land sold during the past year is less
by 252,471 acres than in 1861." The chief cause of this decline, he said, "in official view.....is not accidental or temporary. It is the fact that the best lands of the Crown in
both sections of the Province have already been sold." The
good lands between the Ottawa valley and Lake Huron are,
he says, "composed of small tracts, here and there, separated
from each other by rocky ridges, swamps and lakes, which
render difficult the construction of roads, and interrupt the
continuity of settlement. These unfavorable circumstances
have induced the better class of settlers in Upper Canada to
seek, at the hands of private owners, for lands of better
quality and more desirable location." And in his report for
1865 he says: "The remaining public lands in Canada, from
their general remoteness and their character, are much less
desirable for settlement than those in the valleys of the St.
Lawrence and Ottawa, and of the great lakes." He therefore
recommended that the chief part of the ungranted lands "be
set apart....as a pine region, and no sales be made." This
subject, being one of great importance, has been frequently
referred to in the Canadian Legislature. In 1865, Mr. Mc-
Givern said: "There is not any appreciable quantity of grain-
producing land in the hands of the Government not now
under cultivation in Canada for the benefit of our increasing
population. It is," he said, "a melancholy fact that for the
want of such a country our youth seek homes in a foreign
land." And the Hon. Mr. Mackenzie said in Parliament,
in 1867, that the young men "are now compelled, in
consequence of the limited field for settlement offered in

4

Canada, to seek for homes for themselves in the United States."

And Mr. Russell says: "After all the vacant public lands of the western Peninsula and other parts of Upper Canada, south of the Laurentian formation, were surveyed and sold, or nearly so, the people began to occupy the inferior lands of the Huron and Ottawa Territory, which are in a region of Laurentian formation, at the outline of which settlement had long before, as it were, instinctively stopped." He says there is not fertile land in Canada "to meet our greatly increased native demand;" hence, "It is evident that we cannot attract the immigration we desire to make us a strong people while we have nothing better than that to offer."

Pages might be cited, from the Crown Land, Railway, and other reliable reports, showing that there is no appreciable quantity of fertile land, climatically adapted for settlement, between Red River valley and the Atlantic Ocean. The writer, too, during the last forty years, has had occasion, as a land surveyor and otherwise, to traverse large sections of the country referred to; and can, from personal knowledge, verify the citations made as to large sections of the country.

Perhaps we shall surprise our readers when we assert that there is not as much first-class land in all that part of the Dominion of Canada situated between the Atlantic Ocean and the Red River valley, a distance of over two thousand miles, as is comprised by two of the medium sized States in the fertile belt of the Union. It is now evident that there are no good lands to spare for emigrants in all this vast area. What vacant lands there may be that are fit for tillage, should be left to be filled in from adjacent settlements.

When properly understood, this statement will serve to dissipate some of the dreams, regarded as realities, about a Canadian nationality.

The valley of the River St. Lawrence is the centre from which some suppose a great nationality, containing a hundred millions of people, is to extend east and west. How far this dream may be realized, may be infered from the facts set forth in these pages. It is true, the valley of the St. Lawrence is the front of an immense region; but it is equally true that the fertile part of the Canadian side of the valley is limited to a narrow strip. Guided by a map showing the civil divisions of the country, the reader might conclude from this valley a great empire might spread out, west and north; but when we consider that out of the 185,000,000 of acres of land contained in Ontario and Quebec together, there is only two small areas of first class soils, our hopes of a great future for this country are not bright.

The southern flank of the Laurentian formation is well defined; and the northerly margin of the Notre Dame mountains is equally well marked. Hence, the fertile belt in these two Provinces is limited indeed. The Peninsula of Ontario contains about twelve millions of acres, of which, according . to Sir W. E. Logan, about six millions and a half are highly fertile. And according to the same reliable authority, there is an equal area of good land stretching up the Ottawa valley and down the St. Lawrence." "Both areas," he says, "possess soils of remarkable fertility," and "are endowed with great agricultural capabilities." These two areas together are about equal to the area of Nova Scotia. It is on these two fertile spots that the chief part of the population of Ontario and Quebec reside. Outside of these areas there is comparatively little land in these two Provinces fit for settlement.

Below the city of Quebec there is an area, including both sides of the St. Lawrence, of about eighty millions of acres, on which a population of 269,000 resided in 1861; to this number only forty-one thousand were added in the ten years ending in 1871. Below this city the two mountain ranges, the Laurentides on the north, and the Notre Dame mountains on the south, are close to the river St. Lawrence; and the Gaspé Peninsula, says Sir William, " is a block of table land of about 1,500 feet in height."

About two-thirds of the inhabitants of Ontario reside on the Peninsula; and the remainder, except a few isolated settlers in the mountain region, occupy the narrow valley of the St. Lawrence below the Niagara Falls. Previous to 1860, nearly all the fertile lands of the Crown in these two Provinces, climatically adapted for settlement, had passed into private hands; consequently, for the want of such lands, thousands have been compelled to settle on the inferior lands in the mountain regions. But progress has been very slow. There were sixteen large districts in Ontario in which the total population in 1871 was thirteen thousand less than it was in 1861; and in twenty other districts the total increase was only nine thousand.

And in the Province of Quebec, the progress of settlement in the mountain regions has been remarkably slow. In twenty-two large districts the aggregate population in 1871, was 43,000 less than in 1861; and in fifteen other districts the total increase was only seven thousand souls in ten years. Such is the inhospitable character of the country on the north side of the St. Lawrence, in the Province of Quebec, that the increase in its population in the ten years ending in 1871, did not exceed sixteen thousand persons, exclusive of those in the cities. And the future of this immense region, and also of that below the city of Montreal on the south, is not hopeful.

The Eastern Townships, as far down as the city of Montreal, contain about ten million of acres, which, in 1871, was inhabited by half a million people.

The following table shows the area in square miles of each Province in British North America; and also the population of each for the years named therein, exclusive of Indians:—

| NAME OF PROVINCE. | AREA. SQUARE MILES. | POPULATION. | | |
|---|---|---|---|---|
| | | 1851. | 1861. | 1871. |
| Ontario.......... | 107,780 | 952,004 | 1,396,091 | 1,607,873 |
| Quebec .......... | 193,355 | 890,261 | 1,111,566 | 1,184,528 |
| Nova Scotia...... | 21,731 | 276,854 | 330,857 | 386,134 |
| New Brunswick... | 27,322 | 193,800 | 252,047 | 284,191 |
| P. E. Island...... | 2,100 | 62,678 | 80,857 | 93,698 |
| Newfoundland.... | 42,000 | 101,600 | 122,250 | 146,536 |
| Manitoba........ | 14,000 | .......... | .......... | 12,228 |
| British Columbia, | 356,000 | .......... | .......... | 10,586 |

Including 102,358 Indians, British North America contained 3,833,132 persons in 1871.

This country doubled its white population between 1821 and 1841, and also in the twenty years ending in 1861. Had this increase continued at the same rate, their population would number:—

In 1871............4,701,364    In 1891........... 8,825,000
In 1881............6,441,000    In 1901...........12,090,000

Beginning early in the present century, the population of Ontario increased very rapidly until near the end of the decade ending in 1861, when nearly all the fertile lands of the Crown had been sold. Up to that date emigrants settled in the country in large numbers. Since that time the great body of immigrants passed through on their way to the Western States. And as the inferior lands began to be filled in, the natives of this, and indeed of all the provinces, removed to the States in large numbers.

Had the ratio of increase previous to 1861 continued, Ontario would have contained 2,136,308 inhabitants in 1871; and Quebec 1,422,546. The increase in Ontario between 1851 and 1861 was 414,087, or fifty-three per cent.; in the following decade it fell to less than half that number; and for all British North America, the increase in the decade ending in 1871 was. only 436,120.

The following tabular statement, compiled from the census of 1871, shows the quantity of land owned, occupied and improved, in each of the Provinces named therein:

| PROVINCES. | ACRES OWNED. | ACRES OCCUPIED. | ACRES IMPROVED. |
|---|---|---|---|
| Ontario............ | 19,605,019 | 16,161,676 | 8,833,626 |
| Quebec............. | 17,701,589 | 11,025,786 | 5,703,944 |
| New Brunswick..... | 5,453,962 | 3,827,731 | 1,171,157 |
| Nova Scotia........ | 6,607,459 | 5,031,217 | 1,627,091 |
| Totals ............ | 49,368,029 | 36,046,410 | 17,335,818 |

The above table may afford a base on which to estimate the quantity of land fit for settlement in these four Provinces. It may not be too low an estimate to assume that the quantity of land owned is equal to the total area of the lands, granted and ungranted, which are adapted for settlement. However, if the two fertile areas in the valley of the St. Lawrence only contain about thirteen millions of acres, as estimated by Sir W. E. Logan, the quantity occupied in the Provinces of Ontario and Quebec includes large areas of second and third class soils. Nearly half the lands occupied are improved; and on the improved lands nearly all the inhabitants reside. The quantity of infertile land owned in each Province is comparatively very large. This is obvious to the observer in passing through the country.

In contrasting the progress of the country, the Province of Quebec presents the most discouraging picture to be found in North America. Two and a half centuries have elapsed since the first settlements were made on the banks of the river St. Lawrence. Passing over the long and troublesome period of French domination, down to 1759, when the French Empire in North America, containing at that time about seventy thousand souls, became British territory, we find that it now contains, after a century and a quarter of comparative peace, only about one million and a quarter of inhabitants. Tenacious of home associations, they have long remained within their mountain walls, until necessity compelled hundreds of thousands of them to emigrate to the United States. Their ancient and historical city, Quebec, founded in 1608, remains nearly stationary in regard to progress. In 1861, it contained about sixty thousand inhabitants, and about the same number in 1871.

The city of Montreal has, within the last half century, progressed very rapidly. In 1720 it only contained about 3,000 souls; iu 1851, 57,715; in 1861, 90,000, and in 1871 its population was 107,000; while in material prosperity its increase has been very great. The past and present indicate a prosperous future for this city. These two cities contain about one sixth of the population of the Province of Quebec.

## THE LOWER PROVINCES.

Second only to the Provinces of Ontario and Quebec, are the Gulf Provinces. Except Prince Edward Island, they all bound on the Atlantic Ocean. Their extensive coast lines are indented by a large number of excellent harbors; those on the Atlantic side are navigable at all seasons of the year. Except Newfoundland, these Provinces are traversed in all directions by navigable rivers. Indeed, it is difficult to find a square mile of their surface without lakes or streams. Hence, no part of North America is so well situated for trade, and but few places on the Atlantic seaboard of America contain so many elements of wealth.

The River St. Lawrence, being the outlet for the commerce of the west, adds very largely to the commercial importance of the Lower Provinces. The Straits of Canso, two miles in width, the outlet between Nova Scotia and Newfoundland, and the Straits of Belleisle, afford safe channels of communication for the ships of commerce between the great lakes and the ocean. As the northern passage, however, lets in too much arctic ice for comfort, the propriety of closing it has been suggested. The almost continual floe which enters the Gulf of St. Lawrence through the Straits of Belleisle, has a very injurious effect on the climate, and hence on the agricultural operations of the Lower Provinces, as well as on those of Quebec. The great fields of ice which enter the Gulf and Straits of

Northumberland by means of this passage tends to make winter, in the Provinces bounding on the Gulf, much longer and colder, the warmth of spring later in coming, the frosts of autumn earlier, and the frosts of summer more frequent. There are no mountain barriers to arrest the arctic waves in their course over the Provinces. New Brunswick and P. E. Island, especially, are very flat. And even the forests which once lined their coasts are rapidly disappearing. These climatic difficulties, together with the infertile character of Quebec and a large part of New Brunswick, has made agricultural progress in this large section of country very slow indeed. And, in consequence of this floe, the season of navigation in the Gulf and Straits is much shortened. However, the project of closing the Straits of Belleisle, which are eight and a half miles in width at the narrowest place, and ninety feet in depth, is chimerical.

The fisheries of the Gulf are among the most valuable in the world. And as time passes their great commercial wealth is becoming more and more highly appreciated. Besides the cod, seal and other fisheries of the deep waters of the Gulf, all the chief bays and rivers, and nearly every mile of its coast line, teems with valuable fish.

The Bay of Fundy, with its extensive arms and tributary rivers, has an opposite effect on the climate of the adjoining country from that of the waters of the Gulf. By means of this Bay and its arms the softening influences of the ocean are conveyed over large areas of Nova Scotia and New Brunswick. The surface of this Bay covers an area of about 5,400 square miles. Its easterly arm extends into the heart of Nova Scotia at Truro; and its northerly prolongations, Chignecto and Cumberland Bays, extend easterly to within fifteen miles of the Straits of Northumberland. The tributaries of these opposite waters interlock, and their tides are only six miles and a half apart.

The St. John is by far the most important river emptying into the Atlantic east of the United States. It is about 450 miles in length, and is navigable for long stretches, thus affording a navigable outlet from the interior of New Brunswick and the State of Maine. Some of its chief tributaries interlock with those of the Miramichi and Restigouche of the north.

With map in hand, we find it no easy task to mark out a country containing so limited an area as Nova Scotia, New

Brunswick, and Prince Edward Island, in the aggregate, where the elements of wealth are so valuable; and where the water-courses of commerce are so numerous, capacious and extensive as they are in and adjoining these three Provinces. The only defective link in their water communication is near the line of boundary between Nova Scotia and New Brunswick.

The country between the navigable waters of the Bay of Fundy and those of the Straits of Northumberland has been thoroughly surveyed with a view to the construction of a canal across this Isthmus. Between 1821 and 1874, five different Engineers have explored the country. Although all the surveys, except the last, were very imperfect, still the reports pronounced the country practicable for canal construction.

The number of surveys made by the Governments of the country, and the interest taken in the project by many leading public men, by the press, and Boards of Trade in all the Provinces, from time to time, shew that this is a project of more than ordinary importance to North America.

The most complete survey was made in the years 1872 and 1873, by G. F. Baillairgé, C. E., under the direction of John Page, Esq., Chief Engineer of the Dominion Board of Works.

These opposite waters present a strange phenomenon in physical geography. Mr. Page says: " In Cumberland Basin, the tides rise from 35 to 46 feet over ordinary low water line. Those at the head of Baie Verte range from 5 to 9 feet. At times the water in Cumberland Basin is fully 18¼ feet over that in Baie Verte; whilst at ebb-tide the water in Baie Verte is fully 19¼ feet higher than that in Cumberland Basin." The line recommended by Mr. Page is from Au Lac, Cumberland Bay, through the valleys of the Missiquash and North-west Branch and Main Trunk of the Tidnish river. The distance from shore to shore is 19.25 miles. And the total length for a half-tide canal, susceptible of being extended to full-tide, is 21.22 miles.

Assuming fifty feet below the lowest water of Cumberland Basin as datum, the summit on the line recommended is 105½ feet, being only nine and a half feet above the highest tides of the Basin. There is but little rock on this line, and at the summit the surface is soft mud to a depth of five feet, so that the clay at the highest part of this valley is but slightly elevated above the highest tides of Cumberland Basin, generally known as Cumberland Bay.

"The principal features of the scheme recommended for adoption," says Mr. Page, "are to make the low water level of the canal about the height of the lowest neap tides in Cumberland Basin, or about 85 feet over datum; bottom to be 69 feet over datum, and to make the high water level about two feet under ordinary spring tides, or 88 feet over datum, so as to leave a range of 3 feet for lockage purposes, etc., during neap tides."

The cost of constructing the canal has been variously estimated, according to the capacity, number and size of locks, and mode of feeding the canal. Some engineers have recommended some miles of river navigation, saving largely in the cost of construction. Francis Hall, C. E., estimated the cost at \$298,000, with six locks of 105 feet by 20½ feet, and eight of water on the mitre sills. Thomas Telford, a celebrated English Engineer, examined Mr. Hall's plans, and recommended a canal 30 feet at bottom, and 72 feet at water surface, the width at top to increase according to the depth of cuttings. The depth on the sills to be 13 feet, with locks 40 by 150 feet, estimating the cost at \$685,952. Capt. Crawley advised 9 feet as the depth.

The Canal Commission appointed by the Government reported in 1871 in favor of a canal of 15 feet in depth, 100 feet at bottom, with locks 40 by 270 feet, at a cost of \$3,250,000. In 1873, Messrs. Reefer and Gzowski recommended a half-tide canal on a different line but of the same dimensions, at a cost of \$5,417,000.

In 1873, Mr. Baillairgé, who made a thorough survey of all the routes, and indeed of the Isthmus at large, and who was therefore every way competent to make a proper estimate of the cost, says that a half-tide canal on the Au Lac and Tidnish line, based on Mr. Reefer's project for twelve hours navigation, would cost \$5,650,000.

Believing that a half-tide canal would be sufficient to meet the requirements of the commerce, we therefore leave the consideration of a full-tide canal at a cost of eight and a half millions of dollars, out of the question.

Thus, the Engineers pronounced the project highly feasible, and not costly; and independent public opinion has continued for the last half century to pronounce the construction of a canal across this Isthmus a public necessity.

5

In 1875, the Central Government appointed Commissioners, one from each of the original Provinces of the Dominion, "to investigate the nature and extent of the commercial advantages to be derived from the construction of the Baie Verte Canal."

In the Report, signed by the Chairman, the Hon. John Young, of Montreal, it is said, "The Commissioners have no hesitation in expressing their opinion that it is not in the interest of the Dominion that the proposed canal should be constructed."

One of the Commissioners, J. W. Lawrence, Esq., of the City of St. John, made "A Minority Report," in which he clearly shows that the Baie Verte Canal is a necessity; therefore, it was not the "opinion" of "the Commissioners" that the canal should not be constructed.

However, when we consider the configuration of the habitable part of the Dominion, we need not be surprised at the adverse decision of the majority of the Commissioners. If the Commissioners had been authorised to take evidence at the same time as to the Canadian Pacific railroad, we would be surprised if they found half a dozen persons in favor of constructing it at the public expense. And such is the remoteness of the Peninsula of Ontario from the Atlantic Provinces, that the latter need hardly expect a favorable decision in such matters by the Upper Provinces. And it is to be expected that Mr. Young would feel a much greater interest in completing the navigation of the river St. Lawrence, and that by way of Lake Champlain to New York, than in the construction of the Baie Verte canal. In parts of Nova Scotia remote from the Bay of Fundy and its prolongations, Chignecto and Cumberland Bays, some of the inhabitants are opposed to the construction of the canal, fearing that some trade might be diverted from their coasts. However, New Brunswick, Prince Edward Island, and a large population in Nova Scotia, and many in Quebec, are in favor of the construction of this work.

In the majority report the Commissioners refer generally to the largest sum, over eight millions and a half, as the cost of the canal, and as one of their objections to its construction. In the minority report Mr. Lawrence says: "There are numbers unfriendly to the work at $8,000,000 who would regard it favorably could a canal be constructed for $5,000,000...."

With the aid of the labour saving machines of the day, not unlike those at work deepening the channel through Lake St. Peter, between Quebec and Montreal, lifting up 250 yards per hour of the bed of the St. Lawrence, a sum *greatly less* than $5,000,000 should construct a *full tide canal*, adequate for all the ends of commerce across the Chignecto Isthmus, uniting thereby the waters of the Bay of Fundy and the Gulf of St. Lawrence." He says, "The canal requires no greater depth than the enlarged St. Lawrence and Welland, viz.: 12 feet on water sills."

Both reports, notwithstanding the adverse opinions of the majority, clearly show that the trade which would pass through the canal would be very large.

In another part of this work we have shown that the Dominion Government has been expending immense sums in enlarging the Welland and other canals in the valley of the St. Lawrence, and in the construction of railroads among the mountains and plains of the northwest territories; and still the work of expenditure in those far away regions has hardly commenced.

The Lower Provinces contain about one-fifth the population of the Dominion, and consequently have to pay one-fifth the taxes; it is only fair that they should have a sum equal, at least, to one-fifth the amount to be expended among the mountains on the main land of British Columbia alone. That sum, say $6,500,000, would be more than sufficient to construct the Baie Verte Canal. But those who have represented the Lower Provinces in the Dominion Parliament seem, by the course they have pursued, more willing to sink untold millions on the construction of worthless works, thousands of miles away from us, than expend five millions on the construction of a highly useful work in our midst.

The boundaries of the Lower Provinces are well defined; hence their outlines are easily traced. Exclusive of Newfoundland, their aggregate area is 51,153 square miles, being about equal to the area of North Carolina, and less than that of Georgia, and but a little larger than that of New York or Pennsylvania. Still, these three Provinces contained only three quarters of a million inhabitants in 1871, while the latter State contained three millions and a half in 1870.

We have no doubt had these Provinces united their destiny with the old thirteen Colonies, their population and material

progress would be more than double that of to-day. We cannot doubt that the canal referred to would long ago have
been constructed, by means of which an immense commerce
would be pouring its wealth into these Provinces. It is not
for want of natural wealth that these Provinces have not kept
pace with Pennsylvania in the scale of progress. Though
agricultural capabilities are not so great as those of that State,
they have natural advantages and valuable resources, the
equal of which that State does not possess. Pennsylvania
has coal; so has this cluster of Provinces in great abundance,
and much more advantageously situated. Besides, the Provinces have hundreds of vessels to carry their products to
market; but, unlike Pennsylvania, they have no market for
them. The Provinces contain large bodies of iron ore of the
best quality; copper ore, manganese, and other minerals
highly useful in commerce, exist in many places. Lime stone,
gypsum, and every variety of stone for building, ornamental
and other purposes, are very abundant. And Nova Scotia has
extensive veins of gold-bearing rocks, which annually produce thousands of dollars worth of gold. The area of forest
land in these Provinces is still large, but for the want of a
free market in the States, the products of the forests are not of
much value. And their waters teem with almost every
variety of useful fish.

No country in the world has so large an amount of tonnage
afloat according to population; and the facilities for ship-
building are very great. These Provinces are large producers
of oats, potatoes, hay, butter, beef, cheese, eggs, farm stock,
and other agricultural products. And the country is full of
saw mills as well as wood. And all these products could be
largely increased if there was any encouragement to do so.
But a surplus is a drug. However, we can use what we require at home, and for the want of a free market in the States
we can sell the remainder at half price.

And our loyalty has also become a drug. It is this thing
called loyalty that has kept the Dominion of Canada behind
even a single State of the Union in the scale of progress.

As to how many inhabitants Nova Scotia, New Brunswick
and Prince Edward Island can sustain by the products of their
soils has often been discussed, but with different conclusions.
The aggregate area of fertile land in these three Provinces,
though much larger in proportion than in the Eastern States,

is small compared to the total area of the country. The writer has traversed the Lower Provinces, except Newfoundland, in all directions, and therefore is enabled to form fair conclusions as to the agricultural capabilities of the country. The area of first class soils is very small; that of the second class, on which the chief part of the farming of the country is done, is much larger; but the area of third, and fourth class soils is very large indeed, much larger than has been generally estimated. The country through which the Intercolonial railroad passes, between Rivere du Loup and Moncton, is a fair specimen of the soils of a large part of these Provinces. The continuity of settlement is broken in consequence of the frequent occurrence of large areas of worthless lands.

Though New Brunswick is larger by 2,234,240 acres than the aggregate area of Nova Scotia and Prince Edward Island, still it is doubtful if its agricultural capabilities are greater than those of these two Provinces. Prince Edward Island has no coal; and the coal fields of New Brunswick, though large, cannot be relied on for a supply of fossil fuel. And a large part of Nova Scotia is outside of its coal beds. However, the means of conveyance, by shipping and railroad, are now so great, and the cost so light, that as wood becomes scarce coal can be supplied as a substitute. But as large areas of Nova Scotia and New Brunswick, especially of the latter Province, are only useful for the wood they contain, coal for domestic wants will not be generally used for a long time in the future.

As a basis on which to rest our estimates, though not a correct one, let us take Professor Johnston's Report on the "Agricultural Capabilities of New Brunswick." He says: If this Province "possesses in its mineral resources an available supply of fossil fuel sufficient for its domestic wants, it might sustain in comfort a population approaching to six millions. On the other hand, if wood is to be grown on accessible and economic places its capabilities sink down to the maintenance of three and a half millions of inhabitants."

While we do not apply Wotten's sarcastic definition to Professor Johnston's report, that a man sent to report on the character of a new country is sent to "lie for the benefit of his Sovereign," still we have reason to believe that his report is largely based on unintentional exaggeration. At the time, 1851, when Professor Johnston wrote his report, but little was known of the interior of New Brunswick. The country was not explored as it has been since for railway and other purposes. At that time highly exaggerated notions existed as to the capabilities of the interior of the Province; and also of its mineral resources, as represented in the reports of Dr. Gesner. And the chief part of the replies to Professor Johnston's questions, asking for information relating to the agricultural capabilities of the Province, were made by gentlemen who had settled on the best lands and in the most eligible situations. Hence his estimates were based on partial information.

Already, for the want of suitable lands, emigration to these Provinces has almost ceased; and thousands of the inhabitants of the country are cultivating inferior lands. There is no room in the Lower Provinces for more emigrants. All remaining lands fit for cultivation will be filled in by the inhabitants of the adjoining settlements. However, if emigration from the Provinces to the States continues, thousands of immigrants, possessed of from one thousand to two thousand dollars, will always find homes in these Provinces.

In the ten years ending in 1861 New Brunswick added 58,000 to her population, while in that ending in 1871 it fell to 27,000. In Nova Scotia the increase in these decades was 55,000 and 57,000 respectively. The increase in the first of these decades was largely due to Professor Johnston's flattering report, especially in New Brunswick, and to the fact that the Crown owned large tracts of fertile lands for sale in both Provinces.

Considering fairly all the facts at command, it is a question if the agricultural capabilities of Nova Scotia, New Brunswick and Prince Edward Island are sufficient to sustain in comfort a population of five millions of inhabitants. And in order to obtain such a result, all the first and second-class soils of these Provinces will have to be brought under a very high state of cultivation indeed. Such a state of cultivation as may never take place, unless in commercial union with the United States.

TABULAR STATEMENT SHOWING THE POPULATION AND OTHER STATISTICS OF THE PROVINCES AND STATES NAMED THEREIN. THOSE OF THE FORMER ARE FOR 1871, AND THE LATTER FOR 1870.

| CIVIL DIVISIONS. | POPULATION | WHEAT. Bushels. | BARLEY. Bushels. | OATS. Bushels. | RYE. Bushels. | BUCKWHEAT Bushels. | CORN. Bushels. | HAY. Tons. |
|---|---|---|---|---|---|---|---|---|
| Ontario, Quebec, Nova Scotia, New Brunswick and P. E. Island, 1871, | 3,561,424 | 16,993,365 | 11,672,479 | 45,610,029 | 1,064,358 | 3,801,593 | 3,905,241 | 3,887,000 |
| New York, 1870... | 4,382,759 | 12,178,462 | 7,434,621 | 35,293,625 | 2,478,125 | 3,404,030 | 16,462,825 | 5,614,205 |
| Pennsylvania, " | 3,521,951 | 19,672,967 | 529,562 | 36,478,585 | 3,577,641 | 2,532,173 | 34,702,006 | 2,848,219 |
| Illinois, " | 2,539,891 | 30,128,405 | 2,480,400 | 42,780,851 | 2,456,578 | 168,862 | 129,921,395 | 2,747,339 |
| Ohio, " | 2,665,260 | 27,882,159 | 1,715,221 | 25,347,549 | 846,890 | 180,341 | 67,501,144 | 2,289,565 |
| Michigan, " | 1,184,059 | 16,265,773 | 834,558 | 8,954,466 | 144,508 | 436,755 | 14,086,238 | 1,290,923 |
| Kentucky, " | 1,321,011 | 5,722,704 | 233,486 | 6,620,103 | 1,108,933 | 3,443 | 50,091,006 | 204,399 |
| Iowa, " | 1,194,020 | 29,435,692 | 1,960,779 | 21,005,142 | 505,807 | 109,432 | 68,935,065 | 1,777,339 |
| Wisconsin, " | 1,054,670 | 24,606,344 | 1,645,019 | 20,180,016 | 1,325,294 | 408,897 | 15,033,998 | 1,287,651 |
| California, " | 560,247 | 16,676,702 | 8,783,490 | 1,757,507 | 26,275 | 21,928 | 1,221,222 | 551,773 |
| Minnesota, " | 439,706 | 18,866,073 | 1,032,024 | 10,678,261 | 78,088 | 52,438 | 4,743,117 | 695,053 |
| Missouri, " | 1,721,295 | 14,315,926 | 269,240 | 16,578,313 | 559,532 | 36,252 | 66,031,075 | 615,611 |

TABULAR STATEMENT SHOWING THE POPULATION AND OTHER STATISTICS OF THE PROVINCES AND STATES NAMED THEREIN. THOSE OF THE FORMER ARE FOR 1871, AND THE LATTER FOR 1870:

| CIVIL DIVISIONS. | POTATOES. Bushels. | TOBACCO. Pounds. | BUTTER. Pounds. | HORSES. Number of | CATTLE. Number of. | SHEEP. Number of | SWINE. Number of | IMPROVED LAND. No. of Acres. |
|---|---|---|---|---|---|---|---|---|
| Ontario, Quebec, Nova Scotia, New Brunswick and P. E. I., 1871, | 50,705,913 | 1,505,932 | 75,172,523 | 862,072 | 2,687,274 | 3,302,873 | 1,418,597 | 17,780,921 |
| New York, 1870.... | 28,558,249 | 2,349,798 | 107,147,526 | 856,241 | 2,086,230 | 2,181,578 | 518,251 | 15,627,206 |
| Pennsylvania, " | 13,020,959 | 3,467,539 | 60,834,644 | 611,483 | 1,505,897 | 1,794,301 | 867,548 | 11,515,965 |
| Illinois, " | 11,267,431 | 5,249,274 | 36,083,405 | 1,017,646 | 1,944,573 | 1,568,286 | 2,703,343 | 19,329,952 |
| Ohio, " | 11,423,109 | 18,741,973 | 50,266,372 | 704,664 | 1,521,421 | 4,929,635 | 1,728,968 | 14,469,133 |
| Michigan, " | 10,322,450 | 5,385 | 24,400,185 | 253,070 | 635,134 | 1,985,906 | 417,811 | 5,096,939 |
| Kentucky, " | 3,133,176 | 105,305,869 | 11,874,978 | 351,200 | 812,380 | 936,765 | 1,838,227 | 8,103,850 |
| Iowa, " | 5,949,912 | 71,792 | 27,512,179 | 482,786 | 1,137,045 | 855,493 | 1,353,908 | 9,396,467 |
| Wisconsin, " | 6,648,349 | 960,813 | 22,473,036 | 270,083 | 831,953 | 1,069,282 | 512,778 | 5,899,343 |
| California, " | 2,251,262 | 63,809 | 7,969,744 | 241,146 | 669,280 | 2,768,187 | 444,617 | 6,218,133 |
| Minnesota, " | 1,944,657 | 8,247 | 9,522,010 | 102,678 | 365,241 | 132,343 | 146,473 | 2,322,102 |
| Missouri, " | 4,479,614 | 12,320,483 | 14,455,825 | 545,822 | 1,269,065 | 1,352,001 | 2,306,430 | 9,130,615 |

As the official reports omit agricultural statistics for British Columbia and Manitoba, we have here, in order to make comparisons, omitted the population of these two Provinces, which, in 1871, did not exceed 23,000. The column headed "population" is exclusive of Indians. The column headed "horses" includes all animals of that species. Under the heads, cattle, sheep and swine, calves, spring lambs and sucking pigs are not given in the United States Report from which we copy, but those for the Dominion are included in the Official Report. Hence the value of the farm stock in these three columns for the Dominion is much less than the above figures represent.

The preceding table is very suggestive. A glance at it will fully confirm the opinion held by the writer of these pages, that the Dominion of Canada is not capable of sustaining, by the products of the soil, one quarter of the population which some profess to believe. Thus, after a century's progress, under the patronage and prestige of Great Britain, the aggregate population of the five chief Provinces of Canada is but a few over that of the single State of Pennsylvania, and six hundred and fifty thousand less than New York. And there are several States in the Union, each of which produced more agricultural value in 1870 than the whole Dominion of Canada did in 1871. In these tables we have not referred to the cotton, rice, sugar and other products of the Southern States, but only to the agricultural productions commonly raised in the States and Provinces named in the tables.

The column headed "Improved Land" shows, by the comparative smallness of the areas under cultivation in all the States named, except Illinois, which is the largest producer, that the land is much more productive in those States than it is in the Provinces.

However, the Dominion exceeds the States generally in the production of barley, oats, buckwheat, potatoes and butter; and all, except New York, in the production of hay.

And the production might be largely increased if the Provinces had as free a market in the States as that between State and State of the Union. But, as it is, the farmers of the Dominion are confined to their own limited markets at home, or pay a high duty to the United States—their only natural market; or, on the other hand, send part of their products thousands of miles to Britain and other countries at a great cost.

6

## NEWFOUNDLAND.

Besides being generally veiled with mists, the rocky coasts of this Island present a bold and defiant aspect from the sea. Abrupt cliffs rise from the water and extend inland for many miles. And the interior of the country corresponds, with but few exceptions, with the forbidding appearance of the coasts. Hills, mountains, barrens, swamps, ponds, mossy plains and stunted forests are its chief characteristics. The number of lakes, there called ponds, is almost incredible. They are of all sizes, from half an acre to lakes 25 to 30 miles in length, and 4 or 5 miles in width, and are to be found on the tops of high hills, as well as in the low valleys. It is estimated that more than one third the area of the Province is covered with fresh water. Yet there are no navigable rivers. The streams are mere brooks, flowing through mossy and rocky tracts to the sea. The coast, however, is indented far inland by a great number of spacious harbors, which afford great facilities for commerce. The climate, except at the south, near the Gulf Stream, is not a genial one. The summer season is short, and the winter is intensely cold. This is owing to the two vast streams of water from the Arctic Ocean, Davis Straits and East Greenland currents, which combine and run by its coasts.

As the settlements are generally around the harbors, where fish are prepared for market, but little is generally known as to the interior of the island, more than the general designation that it is unfit for farming operations. There are, however, some fertile tracts of land in the interior, estimated to contain about three millions of acres. But little farming is done: and probably, in consequence of the coldness of the climate, but little can be done with profit.

A variety of useful minerals exist. On the south side of St. George's Bay there are seams of coal of workable thickness. There is also said to be a vein of coal, about a foot and a half in thickness, near the Grand Pond; and indications of coal have been noticed in other places. Copper, lead, iron, nickel and other ores, also marble and slate, are among the treasures of the Island. Large quantities of minerals are annually shipped. In 1877 one copper mine on the north-east shore produced 50,000 tons of ore, worth about $2,500,000. Though Newfoundland cannot take rank among the agricultural communities of this continent, its mineral and piscatory wealth is very great. These two sources will continue to afford remunerative employment for a large population. By means of a large number of spacious bays, or rather arms of the sea, fisheries exist far inside of the general shore line The land is literally surrounded by valuable fisheries. And outside of the coast fisheries lies the Grand Bank, 300 miles in length by 75 in breadth; the Green Bank, 240 miles in length by 120 broad, and the False Bank, and other shoals of less note. These all abound in codfish. The seal fishery is also an important one. The statistics for 1876 show an export of codfish to the number of 1,364,063 quintals. The total value of all the fisheries was $7,847,660. The annual revenue is about $832,000. A railroad through the island, about 360 miles in length, would open up coal and other valuable mineral resources, and some of its best lands for settlement, and afford facilities for procuring timber and lumber.

In summer the coasts of this Island are thronged by fishermen, many of whom leave before winter sets in. France and the United States claim the right, by treaty, to fish on the banks of Newfoundland and dry the fish on the island which they catch on its banks.

Newfoundland has not yet cast in her lot politically with the Canadian Confederation, and, considering the present and prospective debt of the Dominion, it is doubtful if she could gain, at least for some time in the future, by union.

## THE GULF STREAM.

The following extract from Dr. Hartwig's work on "The Polar and Tropical Worlds," is not without interest, especially to those residing on the Atlantic side of North America:—

"This great equatorial current, or rather series of currents, is the marvel of physical geography. Let us follow that of the Atlantic in its long career. Starting on the line of the equator, it flows northwestward along the coast of South America, enters the Caribbean Sea and the Gulf of Mexico, from which it derives the name of the Gulf Stream It passes out through the Straits of Bemini, between Florida and Cuba, a great river 32 miles wide, 2,200 feet deep, flowing at the rate of four miles an hour. Its volume is a thousand times greater than that of the Amazon or the Mississippi, and its banks of cold water are more clearly defined than are those of either of these rivers at flood. So clear is the line of demarcation between the warm water of the river and its cool liquid banks, that a ship sailing along may be half in one and half in the other, and a bucket of water dipped from one side will be twenty degrees cooler than one from the other. Skirting the coast at a distance of about 100 miles, its width is increased and its velocity diminished. Striking the projecting banks of Newfoundland, its course is deflected almost due east, until it arrives at mid-ocean. Here it spreads out like a fan, skirting the shores of Spain, France and Great Britain. It then divides, one branch sweeping around the west coast of Iceland, the other approaching the shores of Norway; and its temporary influence is perceptible in the ameliorated climate of Spitzbergen.

"It is owing to this great ocean river that the temperature of the western shores of Europe is so much higher than that of the eastern shore of America in the same latitudes. Maury estimates that the amount of heat which the Gulf Stream diffuses over the Northern Atlantic in a winter day is sufficient to raise the whole atmosphere which covers France and Great Britain from the freezing point to summer heat. The olives of Spain, the vines of France, the wheat fields of England, and the green expanse of the Emerald Isle, are the gifts of the tropical seas, dispensed through the Gulf Stream." "The stream," he says, "takes eight months to flow from the Gulf of Mexico to the shores of Europe." And, "In the Gulf Stream the warm current is above, the cold below, while on the coast of Japan a cold current from the Sea of Okhatsk runs on the surface, giving rise to a fishery not inferior in magnitude to that caused on the banks of Newfoundland by the cold current of Baffin's Bay."

## THE COUNTRY BETWEEN OTTAWA AND RED RIVER VALLEY.

Going west from Ottawa, this region is eleven hundred miles in length and hundreds of miles in breadth. To those who delight in seeing fertile lands there is not much pleasure to be derived from reading a description of this part of the Dominion; therefore we purpose to be brief. What is generally known as the Laurentian formation occupies nearly the whole length, and the greater part of the breadth of this vast region. Its elevation varies from 1,500, 1,700 to 2,000 feet above the sea. Geologically considered, this is the oldest series of rock formation in the world. Part of it is known as the Huronian formation. Within the folds of this mountainous country there are extensive swamps, and hundreds of lakes and lacustrine streams of every conceivable shape and size. Indeed, Lakes Huron and Superior may be said to lie within the Laurentide region. Some of the other lakes are of considerable size, especially Nipissing, Nepigan, and the Lake of the Woods. The country adjoining the Laurentide mountains on the north is said to be generally swampy. And "from the Lake of the Woods," says S. J. Dawson, "for a distance of twenty-five or thirty miles westward, swamps of great extent, covered with moss and stunted evergreens, are of frequent occurrence. In other sections considerable areas are occupied by marshes or shallow lakes." And Sanford Fleming, C. E., in his Pacific Railway report, says: "For 80 miles immediately east of Red River the general characteristics are a level and, in some parts, swampy country, with ridges of sand and gravel more or less thickly covered with timber; the next 70 miles are rough, broken and rocky."

Agriculturally considered, this is one of the most worthless regions in America. There are here and there some isolated spots of cultivable land within the folds of the mountains. The longest is near Lake Nipissing, north of Lake Huron; and another in the valley of Rainy River, west of Lake Superior. The latter is swampy, but it is said it might be drained. However, in consequence of the northern aspect and great elevation of the country, but few agriculturists will sink their labor in this region while better lands in more eligible situations can be obtained.

The elevation of the country may be estimated from the fact that Lake Superior, the most westerly lake of the St. Lawrence, is 600 feet above the ocean; and the Lake of the Woods, the most westerly lake of note in the Laurentian region, is 1042 feet above the tide. The latter discharges by the river Winnipeg into Lake Winnipeg, of the northwest. At a short distance west of the Lake of the Woods the Laurentian formation tends northwestward, passing around the east shore of Lake Winnipeg in its course to the Arctic Ocean.

The unfortunate position of the Laurentide mountain region, adjoining the St. Lawrence, being so far south, is a complete barrier to the progress of settlement northward, thus limiting the extent of cultivable land in the Dominion between the Gulf of St. Lawrence and Red River to a very small area indeed. Viewing this fact in connection with the future of Canada, one must have faith sufficient to remove mountains to believe that a national Dominion can be erected out of such fractions of habitable country as the Dominion comprises. And if we extend our view westward, across the great desert and arctic slope, and across British Columbia, our faith decreases. Those who boast of Canada as a "great Dominion" might study its physical geography and resources with profit. They would find that much of their boasting is based on a mere dream.

However, there is no country but what is possessed of some advantages. The mineral resources of the Laurentian formation are considerable: copper, iron ore, and other useful minerals, have been discovered, and no doubt others exist. Some mines are now being worked to advantage.

But the most important feature in and adjoining the Laurentian formation is the great extent of water communication. The St. Lawrence is navigable for a distance of 2,400 miles, to Duluth, at the head of Lake Superior The latter point is the terminus eastward of the Northern Pacific railway of the United States. The international boundary follows near the centre of Lakes Ontario, Erie, Huron and Superior, and their connecting channels, to within 180 miles of the head of the latter lake. From Thunder Bay, Lake Superior, the boundary follows a broken line of water and land communication westward into the Lake of the Woods, thence west on the parallel of 49° north latitude to the Pacific. It is 450 miles by land and water between Lake Superior and Red River. The distance is 410 miles by the Canadian Pacific railroad line. Thus one thousand miles of this railroad route is in the Laurentian formation.

In many places throughout this region there are tracts of forest wood which will be useful in the distant future. But, taken as a whole, this immense region of rocky country is one of the great obstacles to the progress of the Dominion.

## THE CANADIAN PLAINS.

Immediately west of the Laurentian region, and near the centre of North America, we enter the great plains. Between the Lake of the Woods and the Rocky Mountains, a distance of 767 miles, the International boundary, the parallel of 49° north, follows nearly in the central water shed of this part of the continent, between two great river systems, and also near the geographical centre between the Gulf of Mexico and the Arctic Ocean. These plains, both north and south of the water-shed, are the most remarkable in the world. The whole comprises an immense desert, extensive prairies, lofty mountains and hills, hundreds of both fresh and salt water lakes, and two of the most extensive and wide-spread river systems on the continent. The waters of the Southern slope discharge, by means of the Mississippi and its affluents, into a tropical sea, the Gulf of Mexico. This slope, and more than two thirds of the Red River valley, belong to the United States.

The Canadian plains, in which the Province of Manitoba and the District of Keewatin lie, generally known as the north-west, slope northward, and are drained by the Red, Assiniboine, Saskatchewan, Athabasca, Peace and Mackenzie rivers, of the Arctic slope of the continent. Thus, by " the unfortunate choice of an astronomical boundary line," the Canadian possessions on this part of the continent are crowded down the Arctic slope, and hemmed in east and west by interminable mountains. The position of the international boundary in the west, as well as in the east, is far north. In the west the Dominion has no claim on the southern slope of the continent. Step by step, as we go west, the manifest destiny of the country seems to be more and more apparent. The Canadian possessions in the northwest are all on the Arctic slope.

The plains, following the international boundary westward, rise by three upward inclined planes or steppes. The lowest is that through which the Red River flows, and is fifty-two miles in breadth, and terminates at the Pembina mountains, long. 98° west. This steppe, or rather valley, rises east and west from the Red River, its average elevation being 900 feet above the sea. Between the prairie in this valley and the Lake of the Woods, " a comparatively small proportion," says G. M. Dawson, "appears to be fit for cultivation." The westerly margin of the Red River valley is defined by a chain of high lands, which extend northerly from the Boundary line a distance of about 350 miles. These highlands are known as the Pembina mountains, Riding mountains (the latter being 1680 feet above the sea), and the Duck and Porcupine mountains, and the Basqua hills, are continuations northerly of the

Pembina range. The easterly margin of this range may have been the shore line of an inland sea, after the waters had subsided below the next higher steppe on the west. The head of the Red River valley is the only place where the water-shed varies to any great extent from the international line.

The second steppe is 250 miles in breadth, and terminates westerly at the Missouri Coteau. The average height of this section is 1600 feet above sea level. The Coteau is about 4,200 feet above the sea, and extends northerly nearly parallel to the westerly margin of the Red River valley. The Thunder Breeding Hills is a prolongation northerly of the Coteau range.

The third steppe extends to the Rocky Mountains, a distance of 465 miles. This great desert has an average elevation of 3,000 feet above the sea. The base of the Rocky Mountains is 4,000 feet above sea level. This plain is entirely devoid of timber, and both the latter steppes are unfit for cultivation. There is, however, says Professor Dawson, a "fertile belt fringing the eastern side of the Rocky Mountains," which, "in the neighborhood of the forty-ninth parallel, is twenty-five miles in width....This fertile region, according to Palliser, and other explorers, narrows somewhat about fifty miles north of the line, but then spreads eastward, while the mountains tend to the west, and includes a great area of fertile country in the vicinity of the North Saskatchewan." But in consequence of its great elevation, 3,500 feet above the sea, high latitude and northern declivity, exposing it to the fury of the cold waves which sweep over the vast treeless deserts lying to the north and east, this belt is not climatically adapted for settlement. Mr. Grant, in his work, "Ocean to Ocean," refers to this tract as "a broad belt along the bases of the Rocky Mountains to the south of Edmonton, two hundred miles long by fifty broad." Measured on the new map issued by the "Dominion Lands Office, Ottawa," in March last, this belt is sixty miles in width on the international boundary; at fifty miles farther north it is 65, and on the parallel of 51 north it is 120 miles in width. North of this parallel it is shown to spread out into the great Peace River region. It is represented on the map as a "superior grain-growing country;" and contains, between the parallels of 49° and 54°, over twenty-one millions of acres, while it is more than probable it does not contain a moiety of this area of cultivable land.

This map reminds us of what Major Emory, of the United States frontier commission, says of some of the maps of the United States:—"Hypothetical geography," he says, "is pushed sufficiently far in the United States. In no other country has it been carried to such a point, or been followed by such disastrous consequences." On "ill-founded information, maps of the whole continent have been engraved and published in the very best style of art, and sent to receive the approbation of Congress and the praise of geographical societies here and abroad."

As Mr. Dawson was geologist and botanist to the Canadian boundary Commission, and as Capt. Palliser spent three years in exploring the country, we prefer their reports to the information furnished by this map.

Indeed, if this fertile belt cannot be cultivated, the Red River prairie, 46 miles in width, is the only tract of any note fit for cultivation on the international line between Lake Superior and the Strait of Georgia, a distance of about 1,600 miles.

Besides the westerly elevation of the plains, there is a gradual dip of this part of the continent northwards, immediately east of the Rocky Mountains. At a distance of 350 miles north of the boundary line, the base of the mountains is about 3,100 feet; and at 200 miles farther northward it is not more than 2,000 feet above sea-level

The chief rivers and lakes of the Arctic slope are very extensive, and overspread large areas. The Red River, which is the most important one in consequence of its southern position, takes its rise in Lake Traverse in the United States, and discharges into Lake Winnipeg in the Province of Manitoba, about ninety miles north of the international boundary. The valley of this river is 315 miles in length, 225 of which is in the United States. The fertile part of the valley is 46 miles wide on the international boundary line. South of this line it gradually becomes narrower to an average of about thirty miles.

Lake Winnipeg extends northerly from the mouth of Red River about 280 miles. Its greatest breadth is 87 miles; its surface covers an area of about 5,140,000 acres, and discharges by Nelson River, 420 miles in length, into Hudson's Bay. Lake Winnipeg is 710 feet above the sea.

Lakes Manitoba and Winnipegosis lie from forty to sixty miles to the west of Lake Winnipeg, and are elevated re-

7

spectively 42 and 60 feet above it. Each are estimated to contain 1,216,000 acres. These and other lakes in this basin cover an aggregate area of 8,896,000 acres. These lakes being in a flat country are not navigable except for small class vessels.

The most extensive tributary of the Red River is the Assiniboine. It enters the main river at Fort Garry, in lat. 49° 52′ north, and long. 96° 53′ west. The Assiniboine takes its rise north of the 52nd parallel, and about 200 miles west of Lake Winnipegos. Its western affluent, the Qu' Appelle, begins in the infertile region, about midway between the Red River and the Rocky Mountains. The lands on each side of the Assiniboine, for a distance of seventy miles from its mouth—to Prairie Portage, are highly fertile. West of the Red River valley this fertile tract is limited by the lake and swampy region on the north and by the arid plains on the south. Above Prairie Portage for full fifty miles the Assiniboine passes through a sandy region. Immediately above this sandy tract there is a large extent of fertile land, the chief part of which is situated north of the parallels of 50° and 51°, and elevated about 1,500 feet above the sea. It lies north of the great desert. In consequence of the sand plains above Prairie Portage and the rapids below, the Assiniboine is only navigable for canoes. The sand plains tend to decrease the volume of water. At one mile and a half above the mouth there are four rapids "of a very serious nature. The aggregate fall of them in four miles is 13.24 feet." It would cost about three quarters of a million dollars to construct a passage through the rapids for small class vessels.

The chief river of the Canadian plains is the Saskatchewan. Its two great arms, the North and South Saskatchewan, take their rise in the eastern slope of the Rocky Mountains. Their numerous tributaries on this slope spread from the international boundary to 54° 30′ north. The waters of the South Branch and of its most extensive tributary, the Red Deer River, run easterly through the treeless desert for a distance of between five and six degrees of longitude, traversing three degrees and a half of latitude. The south branch of the Saskatchewan is about 820 miles in length.

The north branch and its chief affluent, the Battle River, traverse the "Fertile Belt." The North Saskatchewan is 802 miles in length. The junction of the two great branches of this river, known as the Forks, is in 53° 20′ north latitude.

From this point the waters of the main Saskatchewan traverse a distance of 282 miles to Cedar Lake, on their way to Lake Winnipeg, and through the latter lake and Nelson River to Hudson's Bay. Between the Forks and Lake Winnipeg, long. 101° 30′ west, the Saskatchewan extends to the fifty-fourth parallel of latitude. Hence, from the westerly source of the south branch, this river crosses five degrees of latitude and fifteen degrees of longitude.

The navigation of the Saskatchewan is obstructed in places by rapids. In a letter to the Chief Engineer of the Pacific Railway Survey, in 1874, Alfred R. C. Selwyn, Director of the Geological Survey, says: "Between the confluence of the two Saskatchewans there are numerous rapids. At the Grand Rapids the water falls forty-three and a half feet in a distance of two and a half miles." There are also others which impede navigation. "Towing flat-boats or barges, as practiced on Red River, would, I think, be quite impracticable on the Saskatchewan, for the reason that in many places the current is too strong, and in others the available channels between the islands and sand bars or shoals are too narrow and tortuous.". After a careful examination, Mr. Selwyn concluded that the Saskatchewan might be navigated for a short season by steamers of "moderate length, powerful engines, light draft, and as much strength as possible below the water-line." Such is the uncertainty of navigating this river that he recommends that "proper arrangements for warping boats up these rapids in case of necessity should be made in advance. There is," he says, "another very important matter connected with the Saskatchewan navigation which would require careful consideration. I allude to the great scarcity and poor quality, for steam purposes, of the wood." As a substitute he recommends coal, which might be obtained from the seams which are said to exist near the eastern slope of the Rocky Mountains. However, the very uncertain nature of the Saskatchewan navigation, and the remoteness of the coal seams from the treeless regions below, will always tend to make coal a costly article in the lower steppes of the northwest, either for steamboat or domestic use. And this drawback is a serious one where wood is scarce, the quality poor, and the winters long and terribly cold.

Notwithstanding the numerous rapids and shoals which impede the navigation of this river and its branches, there are long stretches capable of being navigated by small class steamers. But the lower part of the main river being far north, nearly as far as the southern part of Hudson's Bay, it is not probable that the Saskatchewan will be used as a line of navigation.

The most important feature in the North Saskatchewan region is that commonly known as the "Fertile Belt." This belt is said to extend from the junction of the two Saskatchewans up the North Branch for a distance of about 400 miles direct, to the base of the Rocky Mountains. It varies in width from seventy to over one hundred miles. This belt, though generally designated a prairie, is traversed by hills and mountains of considerable magnitude, and by numerous and extensive river valleys, which vary in width from half a mile to two miles, and in depth from 200 to 400 feet. This belt lies from 220 to 250 miles north of the United States boundary; also north of the great American desert. Its elevation, though not so high as the desert, is from 1,500 to 3,000 feet above the sea.

The extent of the treeless region on the international line, according to Captain Palliser, is from long. 100° to 114° west; and the apex of the desert reaches "to the 52nd parallel of latitude." And G. M. Dawson says: "On crossing the Pembina River," on the boundary line, about fifty miles west of Red River, "the eastern margin of the great treeless plain is entered on. No woods now. appear except those forming narrow belts along the valleys of streams, and soon even the smaller bushes become rare."

The northerly boundary of the treeless plains observes a course nearly parallel to the general course of the North Saskatchewan, and at a distance varying from twenty to forty miles to the north of it. This treeless region in Canada is estimated to contain about one hundred and twenty-three millions of acres; of this area the sand plains overspread about seventy-three millions of acres. Hence all the Canadian prairies, except the comparatively small tracts in the Red River and Lower Assiniboine valleys, lie north of the desert. The prairies, like the desert, are generally devoid of timber. In consequence of the aridity of the sand plains in summer, the South Branch of the Saskatchewan is not navigable, except for boats.

In the "approximate classification of the Lands" of the Northwest, as represented on the Government map, there is a large belt of land adjoining the desert on the east and north, and extending from the international boundary, near long. 101° to long. 113° W., and lat. 54° N. This tract is designated as "mixed prairie and timber soil; it is rather light, but produces fair crops, good grazing lands," and contains 30,000,000 of acres. This belt includes the chief part of the fertile belt on the North Saskatchewan.

Adjoining this belt of "mixed prairie and timber soil," to the eastward, northward and westward, is an immense region, represented on this map as "generally excellent soil, with abundance of wood and water," and said to be "admirably adapted for the growth of cereals, especially wheat." This belt on the international line at Red River has a breadth of over ninety miles on this map, while the most reliable authorities give the breadth of the fertile part of the Red River valley at only forty-six miles. Northward of the boundary, for a distance of about 700 miles, to lat. 54° and long. 111°, this belt has an average width of about 120 miles. North of this astronomical point the belt spreads out into the Peace River region, where the land is said to be "of extraordinary fertility." But as the Peace River country is elevated from 1,500 to 2,000 feet and upwards above the sea, and lies between 55° and 60° north latitude, we shall leave the question of its adaptation for settlement for the consideration of hypothetical geographers.

One of the great questions in regard to the future of the Northwest is the scarcity of wood.

The country between Ottawa and the Red River valley is known as the wooded region; and also that between the eastern base of the Rocky Mountains and the Pacific Ocean contains much forest timber. But the great desert and prairie region, nearly 800 miles in width, and situated between the wooded regions, is almost devoid of useful timber, or indeed of timber of any kind.

A few years ago the banks of the Red River were generally clothed with timber; but fires and the axes of the settlers have destroyed much of it, and in a few years more the remainder will disappear. The largest area of wood land in this vast region lies about twenty miles north of the international line, near the 98th meridian. It is situated between forty and fifty miles to the north of Fort Garry. The southerly end of this tract varies from twenty to forty miles in width; it extends northward beyond the limits of the Province of Manitoba,

The Assiniboine and its chief tributary, the Qu'Appelle, traverse this region. Mr. Fleming says both banks of these "rivers are densely wooded, but the wood is of no value except for firing, as it is principally aspen and balsam poplar." Some of the mountain slopes are clothed with a similar class of wood. The banks of the lower part of the Assiniboine River, for a distance of about forty miles in a direct line westerly from Red River, are chiefly denuded of timber. Only a few narrow strips of timber lands remain to meet the increasing wants of the inhabitants. The lands in the wooded districts, except on the banks of some of the streams, are generally unfit for cultivation. The chief part of the forest lands are situated north of the great desert and prairies adjoining.

Mr. Fleming says that "fully one half of the line surveyed from Livingston to Edmunton," say 260 miles, "passed through woodland. Poplar is almost the only description of wood found." The railway line in the prairie country is between the parallels of 50° and 53° north; and the chief part of the forest region is north of these parallels. The lands in the wooded region are generally infertile.

As elsewhere shown, the most eligible parts of the Canadian Plains for settlement are the Red River and Lower Assiniboine valleys in Manitoba. "The area of the lowest prairie" in these two valleys, says G. M. Dawson, is about "6,900 square miles, but of this the whole is not at present suited to agriculture. Small swamps are scattered pretty uniformly over its surface, and in some places very large areas of swampy land occur." Mr. Dawson assumes that about one half this area, or about 2,176,000 acres might be taken "as a measure of the possible agricultural capacity of this great valley." And the official report of the surveys of Townships in Manitoba and adjoining country to the north and west show that one hundred and ten townships, 36 square miles each, or 2,534,400 acres, besides parts of other townships, are unfit for settlement. Of the infertile lands, ninety-five townships lie in the wooded region before described. And north of the townships there are extensive areas of mountainous and swampy regions, and other infertile lands, containing in the aggregate several millions of acres. But, strange to say, nearly the whole of these infertile areas, including the Province of Manitoba, except about a million acres, is represented on this map as "generally excellent soil, admirably adapted for the growth of cereals, especially wheat."

It is very difficult, indeed almost impossible, to obtain reliable information with regard to new countries in North America. There is a tendency to exaggeration; and with regard to no part of the country has this tendency been so strong as in describing the plains of the West and Northwest. However careful we have been not to mislead others, we have found it very difficult to avoid being misled ourselves.

The treeless area in the valleys of the Red and Assiniboine Rivers is comparatively large. Its length, from the international boundary northward, is about one hundred miles; and its breadth, in an east and west direction, is from sixty to seventy miles, embracing one hundred and ninety townships, or 4,377,600 acres. With the exception of some narrow strips on the banks of streams, the latter area is devoid of timber. These two valleys are the most eligible for settlement, being farther south and on a much lower elevation than any other part of the Canadian plains.

Between the wooded district above referred to and the Rocky Mountains there is an immense region devoid of timber. Bishop Taché says: "From the 101st meridian up to the Rocky Mountains, a distance of about 900 miles, there is not wherewith to make a substantial road." Between the confluence of the two Saskatchewans and the international boundary the open country is about 300 miles in width, and farther west it is nearly 400 miles wide. Bishop Taché, in his "Sketch of the Northwest of America," page 15, says: "One must travel in the midst of these vast plains and camp out during entire weeks in the midst of these snowy oceans to understand how scarce wood is there, and yet how necessary it is." Referring to the second steppe, he says: "I have read glowing reports upon the plains; they brought out all the advantages; they particularly described the quantity of wood. But, book in hand, I saw the country described, and asked myself, who is the dreamer, the author or the reader. The only woods of any importance on the prairie, that is timber, are the different kinds of poplar, but particularly aspen, and some birch." Since the publication of Mr. Taché's work, in 1868, much wood has been wasted by extensive fires, kindled accidentally or intentionally.

The effect of so vast an area of treeless country on the future
of this northern region cannot fail to be serious; that is, sup-
posing the prairies are otherwise adapted for settlement. At
Red River the want of wood for fuel, building, fencing and
other purposes, has already begun to be felt. Near the eastern base of the Rocky Mountains, and about 700 miles west of
Red River, there is a coal area of considerable extent, but of
inferior quality.

It is estimated that not less than one third the area of a
country should be in the forest state in order to secure the
greatest climatic advantages, and also to afford wood for the
various requirements of society.

In a previous page we have shown that the plains rise along
the international boundary in a well defined step-like form.
To the Report of Progress, 1874, by Sanford Fleming, C. E.,
we are indebted for correct information on the elevation of the
plains on the Canadian Pacific railroad line. Lake Winnipeg,
into which the rivers empty, is 710 feet above the sea, being
one hundred feet higher than that given by previous explorers.
Red River basin is 765, and that of Manitoba 820 feet above sea
level. At Northcote, 220 miles west of Red River, the elevation
is 1,158; at Livingstone, 55 miles farther west, 1,490; at Edmunton, 2,391, and at Jasper House, Rocky Mountains, it is 3,350
feet above the sea. Near the Rocky Mountains the railroad
line is about 300 miles north of the international boundary;
at Red River it is 84 miles north of the boundary.

Another feature in the physical character of the plains is
the immense basins in which the rivers flow. The Red River
being the lowest is an exception. This river at times over-
flows its banks for a distance of ten or fifteen miles from
its mouth upwards. Above that the banks are from twenty to
forty feet above the surface of the water when at its ordinary
level. The valleys of the streams intersected by the railway
line, and in its vicinity, increase in magnitude as we go west.
The valley of the Assiniboine, in the second steppe, is from
150 to 200 feet deep, and from half a mile to one mile in width.
Some of its tributaries are half a mile wide and from 100 to
160 feet in depth. On the third steppe the valleys in which
the streams flow are from 150 to 400 feet or more in depth,
and from half a mile to more than two miles in breadth. The
valley of the South Saskatchewan is 300 feet below the
general level of the plains; that of the Battle River is from

150 to 270 feet below the level of the adjoining country. The streams are small compared to the broad valleys in which they run, and remarkably crooked, frequently winding from side to side of the valleys. There are, besides, a number of trough-like depressions in the plains, some of which are more than one hundred feet in depth, and several hundred in breadth. Hence the cost of railway construction on the Canadian plains cannot fail to be heavy.

But the most remarkable is the Red River valley. Lake Winnipeg is the recipient of all the waters which flow between the 90th and 115th degrees of longitude, and from the 49th for some distance north of the 54th degree of latitude, embracing an area 1,000 miles in length by 370 in breadth; and the Red River, for a distance of nearly 400 miles to the west of the international line, adds to the volume of Lake Winnipeg.

The fertile part of the valley of Red River is 315 miles in length, 225 of which is in the United States. Between Lake Winnipeg and a point 240 miles south, the declivity of the valley is less than one foot in a mile. And east and west from the river the valley rises from ten to twelve feet in a mile.

The lowest summit between the tributaries of the Red River and those of the Mississippi is only 960 feet above the sea. The Laurentide rock formation on the east, and the sand plain on the west of this depression, are each more than a thousand feet above this summit. The surface of the Mississippi River at St. Paul, a distance of about 400 miles from Lake Winnipeg, is about 670 feet above the sea. The areas included by Lakes Winnipeg, Manitoba and Winnipegos, and the adjoining flat lands, including the valleys of the Red River and Lower Assiniboine, may have been an inland sea, the waves of which washed the higher steppes of the west, and the base of the Laurentian region on the east.

The lakes of the Northwest are not deep basins like the great lakes of the St. Lawrence, yet they are navigable for small class vessels; and the Red River is navigable for small class steamers from Lake Winnipeg to the North Pacific Railroad at Breckenridge in the United States, a distance of 310 miles. From the latter place to St. Paul, the capital of Minnesota, is 216 miles by railroad. From Duluth, at the head of Lake Superior, is 242 miles by railroad to Glindon; and from the latter place a railroad is being constructed northward 152 miles to Pembina, at the international boundary. A railroad

is being constructed by the Dominion of Canada between
Pembina and the Canadian Pacific Railroad line, eighty-four
miles. By these routes it is about 455 miles from Winnipeg,
the capital of Manitoba, to the head of Lake Superior.

The summit at the head of Red River being only 250 feet
above Lake Winnipeg, a canal might be constructed, affording a passage for vessels between the heart of North America
and the navigable waters on the Atlantic side of the continent.
Lake Winnipeg is 110 feet above Lake Superior, forty feet
above the Mississippi at St. Paul, 155 above Lake Erie, 128
above Lake Michigan, 448 above Ontario, and 622 feet above
Lake Champlain. Thus the fall from one water basin to
another is nearly gradual from the centre of North America to
the navigable waters of the Atlantic.

But the most important questions connected with the Northwest are those of climate, water, fuel and the grasshopper
plague. In discussing the climatic question we have viewed
the plains as Arctic in their general character. However, it is
difficult to determine the southerly limits of the Arctic regions
in all places. There are local or other influences of a climatic
character which are more or less peculiar to each of the great
regions of the globe. Some countries as far south as 50°, such
as parts of the Province of Quebec, have a decidedly Arctic
character, while other parts of the world, much farther north,
enjoy a remarkably mild temperature. The country lying
north of this continental water-shed has a gradual dip from
the international boundary northward to the Arctic Ocean.
Hence the Canadian plains are exposed to the full fury of the
Arctic waves. There is no mountain range to break their
force; no warm ocean atmosphere to meliorate the climate.
In winter the wind blows en uninterruptidly over a boundless
ocean of snow, beneath which the frost penetrates the ground
to a depth of from five to eight feet. And the great elevation
of even the lowest part of the slope, that of Red River valley
by the rule of allowing three hundred feet in elevation to be
equivalent to a degree of latitude, would give a very cold
winter climate to this slope. The elevated plateaus of Mexico,
7,500 feet above the sea, being near the equator, are not too
high for profitable cultivation.

Captain Palliser says Lake Winnipeg "has the same altitude above the sea level as Lake Superior, viz.: 600 feet."
But according to the Pacific Railway report, which is no

59

doubt correct, its elevation is 110 feet higher than that given by Captain Palliser, which is not favorable. Besides being about 900 feet above the sea, the vast accumulations of ice in the adjoining lakes on the north, do not tend to mitigate the severity of the winter or shorten its duration in the Red River valley. G. M. Dawson says, p. 279, that "the temperature of the Red River country, like that of the prairies generally, depends very closely on the direction and origin of the wind." The valley between Lake Winnipeg and the Mississippi River serves as a passage for the cold winter waves to roll south, and conversely as a passage for the south wind to move northward. Hence the extremes of heat and cold, in the Red River valley are very great; and in summer the heat of the day is often followed by frost at night. Hence the maps showing an equality of climate in the Northwest with places many degrees farther south, are not reliable guides. An isothermal line from Halifax, latitude 44° 39′, is shown to sweep southerly through the United States, and to terminate in latitude 58° north and longitude 115° west. Many other isothermal lines are drawn on paper, showing a great difference in latitude. "These lines," says Bishop Taché, "are fundamentally wrong, for, I repeat, a single night is sufficient to destroy all analogy with the climate to which they refer."

"I am not surprised," says the Bishop, p. 15, "at the impression produced on the tourist while he experiences the real delights of a summer excursion over these plains. Men whose opinions must have weight have, perhaps, occasionally experienced this delightful influence, and have given a preference to the prairies to which they are not entitled in every respect. Here comes the end of August. Already cold is threatening; severe frosts prevent the ripening of cereals and expose them to complete destruction. At other times a similar result may follow drought. We are on the skirts of the desert; its scorching winds rush over the prairie protected by no elevated land. The freezing wind, little less obstructed on its way from the Arctic regions, combats with its violent rival, and the prairie, the scene of this struggle, sees many hurricanes and hail storms, very destructive to the crops. Enormous hailstones have fallen on the prairie; over large districts not only is the hay destroyed, but the soil is as it were harrowed. There often, too often, the desert scuds out its myriads of grasshoppers over the prairie, and carried squadrons and devouring phalanxes that do not hesitate to starve the poor settler. Winter has arrived in the beginning of November, and continues more or less in April, and, Great God! what winter!"

"I have noted," he says, "a common centigrade spirit thermometer every day during ten years. Thrice during that period it has recorded 40° below zero, and it has also thrice marked 40° above, and on one occasion even 43°. During whole months in winter we have a mean temperature of 30° below zero in the mornings, while at midday in summer we have a mean of 30° above in the shade....All the prairie region is subject to....sudden changes, which often cause very great mischief. I have known the whole harvest crop seriously damaged by a severe frost during the night of the 9th and 10th August, although both days had been intensely hot.....Often mercury is frozen during entire weeks."

Few persons had better opportunities of estimating the character of the Northwest than Bishop Taché. During a residence there of twenty-three years he traversed the plains in all directions, both in winter and in summer. These facts, together with his acknowledged ability to form just conclusions, render the foregoing quotations from his work highly valuable. Subsequent writers have confirmed his testimony. Charles Horetzky Esq., in his work, *Canada on the Pacific*, 1874, p. 227, says: "As one might naturally be led to imagine, the climate of the Upper Saskatchewan country is not by any means a genial one, and if my memory serves me aright, I saw the mercury indicate thirty degrees below zero on the morning of the 9th of November, 1871." He says: "The country bordering on the North Saskatchewan, and also a portion of that adjacent to Manitoba, appears from all accounts to suffer fully as much....from the occurrence of early frosts" as that in the vicinity of Lesser Slave Lake, which is 150 miles farther north, where, on the last day of September, he found "the thermometer standing at seventy-five in the shade," while in the night following there was "a sharp frost." At a short distance northward of the Lesser Slave Lake, "the mercury stood at ten degrees below zero" on the 10th November. G. M. Dawson says, p. 302, "The severity of the winter season is certainly one of the greatest disadvantages of the Northwest as an area for settlement."

The extreme of cold in Minnesota and Montana on the south slope is very great, but comparatively of short duration.

The difference between the climate in the northern and southern slopes is very remarkable. The mean temperature at Fort Shaw, one hundred miles south of the international boundary, on the 112th meridian, and at an elevation of about 3,000 feet above the sea, compared with that at Fort Garry, which is fifty miles north of the boundary, and not a thousand feet above the sea, shows that the climatic conditions of these two slopes are very different indeed. The following comparisons are cited from the Report on the Geology and Resources of the region in the vicinity of the 49th parallel, from the Lake of the Woods to the Rocky Mountains, by G. M. Dawson, F. G. S., p. 300:—

| FORT SHAW. | | FORT GARRY. |
|---|---|---|
| November | 39.92 | 14.58 |
| December | 26.75 | 0.58 |
| January | 21.23 | 2.91 |
| February | 30.39 | 2.99 |
| March | 36.58 | 9.00 |

The mean temperature for the other seven months at Fort Shaw was 58.80; and 53.72 at Fort Garry. Thus the difference between their summer climates is not much, while that of the five winter months is very great, the average for Fort Shaw being 31.00, and that of Fort Garry is only 6.00. The long continuance of so low a degree of temperature in the valley of Red River is not a favorable indication of the climate of the fertile belt, which is farther north, and has a general elevation nearly double that at Fort Garry.

In a public lecture delivered in the city of St. John by Edward Jack, C. E., in 1877, the lecturer showed from the meterological journal kept at Fort Garry by the War Department in 1847 and 1848, an unusually mild winter, "that the thermometer there indicated at times as low as 47° below zero, when the officers made bullets of the frozen quicksilver and fired them from their muskets." He showed that, by the reports of the Dominion of Canada for the years 1874 and 1875, the mean temperature of Winnipeg for January was 25° lower, and for February 15° lower than that of Fredericton, New Brunswick, for the same months.

All reports on the character of the plains concur in the fact that the climate there is very severe in winter. Even that very interesting work, " Ocean to Ocean," by Rev. George M. Grant, 1873, acknowledges that the nights of summer are very

cold, and at times frosty. He did not travel there "in winter,
in face of the biting northern blasts which sweep the boundless wastes of these interminable plains with a vigor and
severity almost Arctic in their intensity." He passed quickly
over them in the most pleasant time of the year, when vegetation was "in the full flush of exuberant verdure;" when the
days were hot, and the nights in August "were always cold,"
and frequently frosty. On the night of the 26th there was a
"heavy hoar frost."

In his general observations on the "very narrow strip" of
the plains over which he travelled, Mr. Grant came to the conclusion, p. 176, that "the climate and the soil are favorable,"
and asks, "What about water, fuel, and the summer frosts,
the three points next in importance!"

"A large population," he says, "cannot be expected unless
there is good water, in the form of rivers, lakes, springs or
wells. In many parts of the United States dependence is
placed mainly on rain water, collected in cisterns, but such a
supply is unwholesome, and to it may be attributed much of
their prairie sickness. In connection with this question of
water, the existence of the numerous saline lakes, that has
been again and again noted, forces itself on our attention; the
wonder is that former observers have said so little about them.
Palliser marks them on his map in two places, but they are really
the characteristic feature of the country for hundreds of miles.
In many parts they so completely outnumber the fresh water
lakes, that it is

'Water, water everywhere,
And not a drop to drink.'

Some of them are from five to twenty miles long; others only
pools. Some are so impregnated with salt that crystals of
sulphate of soda are formed on the surface, and a thick white
incrustation is deposited round the shores. Others are brack-
ish, or with a salt taste that is scarcely discernible."

Referring to the country west of the Assiniboine, near the
South Saskatchewan, Professor Selwyn, in his Geological Report, says, "Many of the saline lakes are as much as three,
four or five miles in length, and occasionally from one to two
miles wide." From an eminence, Lump Hill, 400 feet above
the plain, Professor Hind counted "seventeen large lakes;"
and "low ranges of hills can be discerned in several directions....The view extends to the border of the wooded land;

beyond is a treeless prairie. The so called wooded land now
consists of widely separated groves of small aspens, with
willows in the low places. Much of the land on the Lump
Hill is sandy and poor."

Mr. Selwyn says: "In the true prairie region there is very
little water. In some parts of it one may travel for days in
any direction without meeting a stream. Water is much more
plentiful in the half wooded region....The amount of water
which is discharged from the true prairie region within Canadian territory must be very small. During the summer the
South Saskatchewan and its branches appear to lose rather
than gain in volume as they traverse the plains."

It appears from information obtained by Mr. Selwyn from
the inhabitants of Manitoba, that "The flooding of the Red
River on the melting of the snow in spring appears to be
growing less frequent and troublesome....And there is no
doubt the gradual diminution of the water supply of the 'fertile belt' is a matter for serious consideration." He says:
"Travellers who have been accustomed to cross the prairies
from year to year, notice a growing scarcity of water at the
camping places along the trails." However, in the valley of
Red River, near the lakes, wells of moderate depth in some
places supply sufficient water. At Winnipeg, wells "seldom
failed to furnish water at depths not exceeding seventy or
eighty feet." Around Burnside "there is a remarkable area
in which all attempts to obtain water by sinking wells have
proved failures. Several of these had been dug to depths
varying from thirty to eighty feet." In other places in this
vicinity where the wells have passed through sand, good water
has been obtained at a depth of about twenty feet. Farther
west, where the prairies are from 1,500 to 3,000 feet above the
sea, and the rivers from 150 to 400 feet below the surface of
the prairies, it is a question whether good water can be obtained by digging. And the difficulty of obtaining good water
is enhanced in the "fertile belt" region in consequence of the
vast extent of arid plains which bound it on the south. The
great number of salt basins are unfavorable indications also.

Want of water is evidently among the obstacles in the way
of settling the Northwest prairies.

"The question of fuel is next of importance," ("Ocean to
Ocean," page 178), "in a country where the winters are severe,
for corn cannot be grown for fuel in our Northwest as it has

been on the prairies of Illinois. At present, on account of the destructive prairie fires for successive years, there is little wood except along the rivers and creeks, and on some of the hills, until we go back to the continuous forest on the north, to within two hundred miles of the Rocky Mountains." Near the wooded region, east of the Rocky Mountains, "We have the most extensive, perhaps the finest, coal fields in the world." And Mr. Dawson, in the work before referred to, says, p. 18: "It is likely....that a trough or series of more or less isolated basins of lignite and coal-bearing strata follows near the eastern base of the mountains the whole way to the Arctic Sea." These coal beds extend far south into the United States.

From the work "Canada on the Pacific," p. 202, we again cite: "The scarcity of wood and water are insuperable obstacles in the way of successful and permanent settlement. It is true that occasionally small copses of poplar (the trees rarely exceeding eight inches in diameter) are met with; nevertheless, the extent of wooded compared with prairie land is so disproportionate, that the quantity of wood required by a widely scattered community of settlers would suffice to clear off all the available timber in a very few years."

"On the score of fuel, it might be urged that the coal which underlies a great extent of the Upper Saskatchewan country, may offer a good substitute for wood, and be used to advantage. There is no doubt that coal, in quantity enormous, but in quality perhaps doubtful, is to be found, especially west of Fort Pitt; but those who seek these regions with a view to settlement cannot be expected to turn all their attention, and devote all their energies towards the painful and laborious extraction from the bowels of the earth of the wherewith to keep body and soul together during the long and severe winters which are the rule, when the thermometer often sinks to 40° below zero." So far as known, the coal of this region is generally of an inferior quality.

After discussing the questions of water and fuel, Mr. Grant says: "The remaining difficulty is the recurrence of summer frosts. In many localities these are dreaded more than anything else. At one place in June or July; at another in August, sharp frosts have nipped the grain, and sometimes even the potatoes.

It is often adduced as proof of the mildness of the climate of the plains that horses live in the open air all winter,

Bishop Taché says: "This circumstance, so remarkable to those unaccustomed to this country, instead of proving the mildness of the climate, proves the constancy of cold. Not only does snow not melt in winter, but it does not even soften, thus it does not become icy, nor acquire what is well known in Canada as 'crust.' The horse, by pawing, can easily remove the covering of the snow from off the grass and feed, which would be impossible were the snow to harden....The horse, though an animal of a milder climate, nevertheless withstands the lowest temperature. Surprise at seeing horses wintering in the open air is nothing more than what Europeans experience on seeing Canadian horses, after long journeys, standing out in the cold for hours together without suffering in the least. The fact, then, that horses can live without stabling does not prove the mildness of the climate, but simply the abundance and superiority of the immense pasturages left for their use. This indeed is the unquestionable advantage of the prairie country." Speaking of the horses of the Northwest, "Those animals," says Horetzky, "were of all shades of color, and no two were alike in size. They were of the hardy little breed peculiar to the Saskatchewan country, and though not much to look at were possessed of qualities of endurance hardly to be expected from animals of their appearance."

G. M. Dawson says: "I am aware that in this region horses and cattle are at present frequently allowed during the winter to feed themselves as best they may. They generally survive, and often do not look much the worse for their hard treatment; but this haphazard plan will not find favor with careful farmers. In the Red River country animals to which proper attention is shown require additional food to be supplied to them, either in the form of hay or oats, for at least six months in the year."

Were it not for the terrible winters, summer frosts, the grasshopper plague, and the difficulty of procuring wood and water, we would advise those in search of homesteads to be guided in their selection by the Rev. Mr. Grant's concluding remarks: "Looking fairly at all the facts, admitting all the difficulties—and what country has not its own drawbacks—it is impossible to avoid the conclusion that we have a great and fertile Northwest, a thousand miles long and from one to four hundred miles broad, capable of containing a population

9

of millions. It is a fair land; rich in furs and fish, in treasures of the forest, the field and the mine; seamed by navigable rivers, interlaced by numerous creeks, and beautified
with a thousand lakes; broken by swelling uplands, wooded
hill-sides, and bold ridges; and protected on its exposed sides
by a great desert or by giant mountains. The air is pure, dry
and bracing all the year round, giving promise of health and
strength of body and length of days. Here we have a home
for our own surplus population and for the stream of emigration that runs from northern and central Europe to America.
Let it be opened up to the world by rail and steamboat, and in
an incredibly short time the present gap between Manitoba
and British Columbia will be filled up, and a continuous line
of loyal Provinces extend from the Atlantic to the Pacific."

Locusts, or grasshoppers, are the plague of the west and
northwest. The great sand plains adjoining the Rocky
Mountains is their native place. In the Dominion these plains
lie to the westward of the Red River valley, and to the south
of the Assiniboine and Saskatchewan prairies; hence these
prairies are fully exposed to their ravages. Our allotted space
will not permit our entering upon this subject at length,
though it is one of great importance to the States and Territories adjoining the great desert.

It is not easy to obtain reliable information in regard to the
effects of the grasshopper plague on the Canadian plains.
Some writers simply refer to the subject; others gloss it over
as one of trifling importance. The Rev. Mr. Grant says in
regard to Manitoba, they "have proved a plague only two or
three times in half a century." And the Hon. James Trow,
M. P., in his "Letters" relating to "Manitoba and Northwest
Territories," which were republished by the Department of
Agriculture of Canada in 1878, says, p. 23: "Manitoba had an
immunity of 37 years, from 1820 to 1857; not a single grasshopper was in the country during that long period. In 1873,
1874 and 1875, crops were partially destroyed, more particularly in 1874, but none have since appeared, and in all probability may not again for half a century."

Alex. Begg, in his "Guide to Manitoba," p. 48, says: "In
1819 they destroyed the crops, and for three successive years
the hopes of the husbandman." On page 103, he says: "For
four years," dating from the end of 1875, "the country has
been swept by grasshoppers,"

The following statistics are gleaned from the Geological report of G. M. Dawson, who has devoted much time to the subject:—

In 1818, six years after the foundation of Lord Selkirk's colony, the locusts destroyed nearly everything but the wheat crop, which partly escaped, being nearly ripe. In the following spring all the crops were destroyed. In the two following years the crops suffered greatly from their ravages. Mr. Dawson says: " the next recorded incursion is that of 1857;" but too late in the season to do great damage, " but eggs were deposited, and in 1858 all the young grain was devoured. In 1864 they again appeared, and left their eggs, but neither the adults nor the young of 1865 were sufficiently numerous or wide-spread to do much damage. In 1867 numerous swarms poured in, but did little injury, the crops being too far advanced; their progeny in the ensuing spring, however, devoured everything, causing a famine. They again appeared in 1869, the young in 1870 doing much harm. In 1872 fresh swarms arrived, but as usual too late to do much damage to wheat. Eggs were left in abundance in the northern part of the Province, and in the following spring the farmers over considerable districts did not sow. In 1874 winged swarms again came in from the west, arriving earlier than usual, and inflicting great injury on the crops in some districts. Eggs were deposited in almost all parts of the Province, and the result has yet to be seen." From other sources of information we learned that great damage was done in 1875. According to the official report for the year ending June, 1876, this little colony of less than thirty thousand people imported flour and grain, nearly all from the United States, to the value of $236,-878; and only exported three barrels of flour.

The Canadian Government had to assist the inhabitants of the Northwest repeatedly in consequence of the complete destruction of the crops by locusts. In the correspondence in regard to the transfer of the Northwest to the Dominion, it is stated that "from 12,000 to 15,000 souls were in imminent danger of starvation during the winter " of 1869.

An important question arises in connection with the grasshopper plague. Will their ravages become more frequent as cultivation becomes more continuous? If their native place is the high arid desert it is a question whether human agency can do anything to destroy or in any way lessen their number.

So important indeed is the matter that some of the States which are subject to their ravages have enacted laws with a view to the destruction of the locusts. Their visits seem to have been more frequent and their effects more destructive at Red River within the last ten or fifteen years than formerly.

*Area of the Canadian Prairies.*—The Saskatchewan and other tributaries of Lake Winnipeg are said to drain about 280,000 square miles, or 179,200,000 acres. This area comprises immense tracts of desert prairie, mountains, swamps and river beds. It is not easy to arrive at a reliable estimate of the extent of fertile land within these bounds. In no part of the world has that kind of patriotism which substitutes fiction for truth been more fully exercised than with regard to the extent of the Canadian prairies, and their adaptation for settlement. There is no doubt, however, that the area of fertile lands is very great. Even far north of the fifty-fourth parallel, it is said there are extensive tracts of rich prairie lands. But the great question is, how much of even the most southerly prairies of the Dominion are climatically and otherwise adapted for settlement?

Captain Palliser and Professor Hind estimated the aggregate area of prairie lands lying south of the parallel of 54°, at about 41,000,000 of acres, about equal to the area of fertile lands in the State of California. Of this, says A. J. Russell, C. E., in his work on the Northwest: "The one-third of the fertile region estimated by Captain Palliser as being" fit for cultivation, "is a very fair proportion; the other two-thirds, no doubt, are requiring drainage or particular clearing." "As for the 80,000 square miles," or 51,200,000 acres, "which Captain Palliser designates as the least valuable part of the prairie country, it will no doubt, as he says, 'be forever comparatively useless,' with the exception of such tracts as the Cypree Mountains, and others where there is good grass, with wood and water." Bishop Taché says, p. 13: "It is difficult to give, even approximately, the area of these prairies. I reckon them as being about equal to the desert country, that it 60,000 square miles," or 38,400,000 acres. At page 19 he says: "At the risk of appearing to be unreasonably retrograde, I dare positively affirm that not more than one-half of the area of the prairie within the limits I have ascribed to it, or within the region usually called the Fertile Belt of the Northern

Department, is fit for settlement, and this half has not all the advantages attributed to it."

From all the facts at command, it is obvious that a value has been ascribed to the Canadian Northwest which it does not possess. Nature in this region assumes conditions at variance with those of other places where agriculture is pursued advantageously. On these plains there is a long and continual struggle in winter between life and frost, and but little wood. In summer frosts are frequent. Besides, there is a "long period of the year," says G. M. Dawson, "during which out door agricultural work is impossible." And in his "Guide to Manitoba," Alex, Begg, Esq., says, p. 101, that "it should be remembered that when the winter comes employment in Manitoba ceases, and the immigrant who may have worked hard in fencing and breaking land, raising a house for himself and buildings for his animals, has a long winter to pull through."

Balancing the advantages and disadvantages connected with settlement in the Northwest plains, it is impossible to overlook the many serious obstacles in the way. Whatever the wants of the American people may be in the distant future as regards room for settlement, it is not probable that many persons will prefer a home on these dreary plains when good lands can be obtained in the genial climes of the continent. Hence but few settlements will be made, at least for a long time to come, outside the Red River and Lower Assiniboine valleys. These two valleys are highly productive. The same geological causes which have endowed the Western States and California with such vast agricultural resources as they possess, have contributed to the richness of these two valleys, especially that of the Red River. Professor Hind assumes "that the prairies of Red River, and of the Assiniboine east of Prairie Portage, contain an available area of 1,500,000 acres of fertile soil;" and G. M. Dawson says these valleys contain about 2,176,000 acres of fertile land. And the United States Agricultural Report for 1868, p. 457, estimates the United States part of the Red River valley at 11,500,000 acres. The latter is, no doubt, too large an estimate. Out of the best portion of the fertile lands of the Dominion, there is reserved 1,400,000 acres for the children of the half-breeds, whose parents had already had a title to most of the land along the Red and Assiniboine rivers, extending back on both sides of these streams a distance of about four miles. In addition, the Hudson Bay Company have reserved to them over 700,000 acres. Thus nearly all the fertile part of these valleys in the Dominion is already disposed of.


### BRITISH COLUMBIA.

This is emphatically a mountainous region, and consequently unfit to be classed among agricultural countries. Still it has not been an entire blank in history. At times two governments existed, one on Vancouver Island, the other on the mainland. The population of both sections did not exceed in the aggregate that of a small country town; even in 1871 the total population numbered only 10,586, and the increase since that time has been remarkably slow. The two colonies are united—a province of the Dominion of Canada. The discovery of gold in the sands of Fraser River and other places, gave rise to high expectations. In 1858, about twenty thousand persons, chiefly from California, rushed to the country in search of gold. The greater part returned much disappointed. In 1859 a dispute arose with the United States as to the possession of San Juan Island. That power claimed it by virtue of the Oregon treaty, and sent a company of soldiers to take possession of it. England claimed it, and sent her ships of war to defend it, whereupon the United States force was largely augmented. For a few weeks war was imminent. A joint occupation of San Juan was agreed to pending the settlement of the dispute, which settlement was effected through the Emperor of Germany as the arbitrator, whose decision was in favor of the United States. Thus the British claims, and the control of the channels leading to the inner waters of British Columbia, passed into the hands of a foreign power.

British Columbia has occupied an important page in Canadian history also. She entered the union in 1873, on condition that the Dominion Government would connect the Pacific Ocean by railroad with the railroads of the Ottawa region in ten years from that date. The engineering difficulties being such that this could not be done, hence the terms of union have not been all fulfilled by the Canadian Government. The Local Government of British Columbia appealed to the Colonial office in London, complaining of their non-fulfilment. The time for the completion of the work has been extended, and will have to be further extended before this vast work can be accomplished by the Dominion.

For a description of the physical character of British Columbia we are indebted chiefly to the reports of Mr. Fleming, the Chief Engineer of the Canadian Pacific Railroad survey. On this part of the Pacific coast there is a great number of Islands, including the San Juan group. "Vancouver Island is the most southerly and the largest. Its extreme lengh is about 280 miles; it extends northerly and westerly from the Straits of San Juan de Fuca in a parallel direction to the mainland. One hundred and thirty miles northerly, and slightly westerly from Vancouver Island, the Queen Charlotte Islands begin, a group of three islands, separated by narrow channels and extending along the shore nearly 200 miles.

"These islands have distinct mountain ranges of their own, with central peaks rising up from 6,000 to 7,000 feet above the sea....The exposed coasts of these islands are characterized by bold rocky headlands, between which deep, narrow sheltered inlets pierce to the heart of the mountains. From the open sea the mountains present a lofty serrated outline."

Dr. Hector says: "The southern part of Vancouver Island, where the town of Victoria is built, is composed of metamorphic rocks, with occasional beds of crystalline limestone. This district, and also the central portion of the island is, as may be expected from the formation, everywhere hilly, and even mountainous, with only limited patches of fertile soil in the valleys. However, the scanty soil on the rocky hills supports a fine growth of timber, so that they are almost invariably wooded to their summits. In the immediate neighbourhood of Victoria there is, nevertheless, a good deal of fine open land, dotted with small oak trees."

Referring to the mountain regions, Mr. Fleming says: "The coast chain, generally called in British Columbia the Cascade range, and southwards the Sierra Nevada, runs generally parallel to the coast; although south of the 49th parallel, in California and other States there are intervals of broad plateaus between the Pacific Ocean and the foot of the mountain slopes; but northwards of the mouth of the Fraser River, along the whole coast of the main land of British Columbia, the mountain slopes come sheer down to the waters of the Pacific."

Between the Cascade range and the Rocky Mountain chain "is an elevated undulating plateau, ranging from three

thousand to four thousand five hundred feet above the level of
the sea. This is much broken by lakes and spurs from the
main mountain chains and inferior parallel ranges, and by deep
valleys, through which flow the rivers on their course to the
Pacific Ocean.

"The breadth of the coast chain on the lines which we have
surveyed is from 100 to 120 miles from the inlets of the Pacific
coast on the west to the foot of its eastern slope. The western
slope is indented with numerous fiords or deep water arms of
the sea, running 30 to 60 miles into the mountain chain, and
the main ranges are a chaos of bold, rugged mountains of
bare rock, rising abruptly and terminating in irregular masses
of snow-capped peaks, from 6,000 to 10,000 feet above the
level of the ocean." That is, the average breadth of the Cascade range is about 150 miles. "On the eastern slope of this
chain, and extending on the plateau between it and the Rocky
Mountains, is a belt varying in breadth, but probably averag-
ing over one hundred and twenty miles, which is sheltered
from the rain clouds coming from the west by the great elevation of the Cascade Mountains; on this but very little rain falls,
and there is consequently scarcely any underbrush, and the
larger trees, chiefly firs, are thinly scattered, singly or in
clumps, giving the whole country a park-like appearance.

"On the western slope of the Rocky Mountains, and in the
high plateau between them and the coast chain, several large
rivers have their sources. Those flowing westward or southwest have cut their way through depressions in the coast chain
to the Pacific Ocean."

"But unfortunately," says Mr. Fleming, "though the rivers
(especially the Fraser) descend with tolerable uniformity, the
valleys in British Columbia, everywhere narrow, do not leave
much margin between the rivers and the foot of the slopes of
the hills or high plains that bound them, and as the rivers
roll onward to the ocean, cutting deeper into the earth, this
margin becomes more and more contracted, till, on entering the
foot hills of the Cascade chain, it entirely disappears, except
where depressions have been made by lateral streams, and the
valley becomes a mere gorge or trough."

From the foregoing quotations the reader might conclude
that there is no land fit for tillage on the main land of British
Columbia, but Dr. Hector says the mountains "retire along
the north shore of Burrard's Inlet to the southeast, so as to be

sixty miles inland at where the boundary meets them, thus leaving a very heavily timbered tract, which forms the only level country in British Columbia east of the Cascade range."

Lord Dufferin described the front of British Columbia as "presenting at every turn an ever shifting combination of rock, verdure, forest, glacier and snow-capped mountain of unrivalled grandeur and beauty."

The climate of the southern part of Vancouver Island is moderate, and well adapted to agricultural operations. North of the island the climate is remarkably wet. The high mountain summits, in all directions, are capped with snow at all times, and in the valleys between, which are remarkably narrow, snow falls to great depths, and in places lies the greater part of the year.

It is clearly obvious that farming will never be of much note on the Pacific coast of the Dominion, nor on the United States side adjoining. There is, however, considerable timber on the mountain slopes, if it could be brought to the sea-board. There is abundance of useful fish in the inner waters; there is excellent coal on the inner coast of Vancouver Island; and iron ore, and more or less gold and other minerals exist in the country. .

But similar resources exist in great abundance on the United States side of the boundary also. Hence it is difficult to find a market, except in far distant Asia. For these reasons British Columbia, though of immense area, is not likely to contain half as many people, or be worth half as much to the Dominion as that of little Prince Edward Island in the East.

## EXAGGERATIONS.

In North America there is a constant tendency to overvalue the resources of the country, which has often led to expectations and awakened hopes which have not been and never can be realized.

Countries in America, however, are on a large scale, and generally possess vast resources; it is, therefore, natural for their pioneers to be proud, or even boastful. But fiction should not be substituted for truth.

In the United States this kind of patriotism began with the nation and increased with its growth. Regions of worthless lands have been sounded abroad as the most valuable. Hence thousands of emigrants have been seriously disappointed, while their money has been paid to speculators or other interested parties.

In the Dominion of Canada, boasting is a vice of more recent date, and is chiefly confined to a few of the officials of the country, who are able, in this respect, to compete with their Republican neighbors. They boast of Canada being a "great Dominion," of having immense regions of fertile lands within the range of climatic adaptation. In previous pages we have shown that a value has been set on the country for settlement which it does not possess.

No part of America has suffered so much by means of exaggerated statements as British North America. Hundreds of thousands of copies in the aggregate of papers, pamphlets and highly colored maps, setting forth advantages as to climate, soil and extent of country fit for settlement, which the Dominion does not possess, have been circulated in the British Isles. Hence thousands of immigrants have been disappointed, and consequently removed to the United States, after a short residence in this country. Millions of dollars, even as high as half a million a year for some years, have been spent in this way, a large part of which has aided in filling in the United States. And at no period have exaggerations been so rife as during the last ten years. The most worthless regions have been lauded both at home and abroad. The fact of keeping agents pleading at the thoroughfares of the British Isles for people to emigrate to this country, is itself sufficient to raise doubts as to the fitness of the country for settlement. If the Dominion is what it is represented to be by immigration agents and others, its present inhabitants ought to be able to induce friends and relatives at home to emigrate to this country, independent of the present costly system.

The following citations, out of many others which might be adduced, will suffice to show to what extent exaggerations have been indulged in.

British North America, said the Hon. Joseph Howe, contains "4,000,000 square miles. The United States has not so much. All Europe with its family of nations is smaller by ninety-two thousand square miles." Again he said: "I often smile when I hear some vain-glorious Republican exclaiming

> ' No pent up Utica contracts our powers,
> The whole unbounded continent is ours !'

forgetting that the largest portion does not belong to him at all, but to us, the men of the North, whose descendants will control its destinies forever. The whole globe contains but thirty-seven millions square miles. We North Americans, living under the British flag, have one ninth of the whole, and this ought to give ample room for the accommodation and support of a countless population."

And in reference to Upper and Lower Canada, Mr. Howe said: "The Province of Canada is as large as Great Britain, France and Austria put together, and will, if ever peopled, sustain a population of fifty millions." Probably Mr. Howe did not know that less than one sixth part of the area of the two Canadas is fit for settlement.

But the Hon. Mr. Mackenzie is a full match for Mr. Howe. We quote from a book entitled his "Speeches during his recent visit to Scotland." In page 21 he says: "Our prairie land alone, west of the great lakes, upon which we have but just entered, extends for a distance of nearly 900 miles, with a width of at least 300 miles, and that there is forest land for many hundreds of miles to the north and west." But the following crowns the climax: "We have," he said, "between the Rocky Mountains and the Lake of the Woods, 1,000 miles territory from east to west, by 500 miles from south to north, fit for settlement." That is, 500,000 square miles or three hundred and twenty millions of acres, situated north of the parallel of 49° and the great sand plains, *fit for settlement*. As to what reliance can be placed in this statement, we refer the reader to our description of the Northwest in previous pages.

We might be permitted, however, to ask: Were these statements made before British audiences with a view of inducing British capitalists to lend us more money, and the British nation to defend the Dominion of Canada in the event of war with the United States?

Whatever may have been the object in making such unwarrantable statements, we leave the reader to judge. However, by thus playing on the credulity of his audiences, some were led to believe these statements to be literally correct. For example, the *Dundee News*, in its comments on the utterances of Mr. Mackenzie, said Canada contained a "splendid stretch of prairie and forest land, on which all the inhabitants of Europe could easily settle;" and that "no one could help thinking what a glorious theatre is here prepared for one of the proudest scenes that has ever been enacted in the world's drama." Europe contains about three hundred millions of inhabitants, all of which, we are told, "could easily settle" in the Arctic slope of North America. Thus, in imagination, the then Premier carried his audience far down this slope; indeed, it is not easy to say at what point northward he terminated his measurements; in fact, he might have extended them much nearer to the Pole with equal propriety. If the *Dundee News* had been aware that the boundaries of the Premier's domain included hundreds of millions of acres comprising deserts, swamps, mountain regions, and other infertile lands, besides millions of acres covered by water; had the *News* been informed that Canada adjoining the international boundary, between the Lake of the Woods and the Rocky Mountains, a distance of 800 miles, and for a breadth of at least four degrees of latitude, does not contain twenty millions of acres of land adapted for settlement, as far as regards the soil, and that hundreds of millions of acres of the Canadian Northwest are devoid of wood, that water is scarce over large areas, and that the winter climate is a terrible one, it might have concluded that the population of Europe would find but a dreary home on the Canadian plains.

"I look upon Canada," said the late Premier, "as a country peculiarly favored as to climate, as to soil, as to the feelings of the people, as being the home of a brave, a generous, and a powerful nation, and one yet destined to play an important part in the history of the world....I am aware that many people emigrate to Canada, and that some return with feelings of disappointment. They do not think the river St. Lawrence is as big as has been said. They do not think that the Falls of Niagara quite so large as has been stated. They do not think the lakes are so vast as has been represented to them." This is concealing the truth under a shower of words. He might as well have said that some return with feelings of disappointment because there are no whales in Lake Superior. In 1867 Mr. Mackenzie gave the true reason why many have removed from Canada when he said that Canadians "are now compelled, in consequence of the limited field for settlement offered in Canada, to seek for homes for themselves in the United States."

Speaking of the future population of this country, he said: " Perhaps not in my lifetime, but possibly in the life of my immediate successors, a larger population will inhabit the British portions of North America than now inhabit the British Isles." He did not name the precise time when British North America will contain thirty-two millions of people. It is probable that Mr. Mackenzie's successors will have passed away long before this country will contain one-fourth of the present population of the British Isles.

This prophecy as to the future population of this country reminds us of one of a similar nature by the Canadian delegates to Great Britain in 1862. In an official dispatch asking Imperial aid to construct the Intercolonial railroad, they told the British Government that the population of these Provinces " will numerate at least twelve to fifteen millions in twenty-five years." When the half of this period expired, the total population of the Dominion and Newfoundland together did not exceed four millions, including 100,000 Indians. And in 1887, when the term will expire, there is no probability that the population of British North America will exceed five millions.

The fact is, we have been estimating our future millions without an adequate knowledge, or with exaggerated notions of the resources of the country. Of this nature is the assertion of the Hon. Joseph Howe, that the two Canadas could sustain a population of fifty millions; and that of Sir E. P. Taché, who said in 1865, "That in less than half a century

Canada," that is Upper and Lower, Canada, " would embrace a population equal to that of the large empires of the old world." Fully one fourth of this time has already expired, yet the population of these two Provinces is only about three millions and a quarter. At their present rate of increase, and we have no reason to believe it will be greater in the future, many half centuries will expire before the aggregate population of Ontario and Quebec will be equal to half that of one of the small empires of the old world.

In 1874 John Boyd, Esq., in a lecture in the city of St. John, is reported to have said, in reference to the Dominion, that " We are great in territory—the largest country in the world, excepting Russia; we have room and homes for 100,000,000 population." And in a public lecture in the same city in 1877, by the Rev. James Bennet, the lecturer assumed that " at no distant date," the Dominion will contain " ten to twenty times" its present population. He did not name the exact time when this prophecy shall be fulfilled, but he asked: " Would it be too much to say that in the future this Dominion may contain 100,000,000 souls? I am not," he said, "a good actuary to calculate how long it would take at the present rate of progress, which is about $1\frac{1}{4}$ per cent. per annum, to reach that figure, but assuming the same rate of progress as the United States during the first century, we—since we have about the same population to begin with which they had—should be 40,000,000 during the next century, and, and in another century thereafter we should be 100,000,000. Well, we do not calculate to grow quite so fast. It may be thought that even 100,000,000 during the next 200 years is too much to count upon." With reference to the sections of the Dominion which he considered capable of sustaining many people, he said: " We have....irrefragable testimony that the prairie grounds in our Northwest are more productive than any soils in the United States, with the exception of those in California." The area of the prairies he estimated at " 480,000 square miles, or equal to ten States, each the size of New York." But the most remarkable statement is that " nearly the whole of this territory is as well or better fitted for settlement than the best parts of Ontario, New Brunswick or Nova Scotia, is ready for the plough, produces wheat and other cereals, and all kinds of vegetables, of the best quality. There is no doubt," he says, " that this district alone is capable of sustaining a population of 80,000,000 people,"

Thus, according to Mr. Bennet, we have upwards of three hundred millions of acres of prairie grounds in the Canadian Northwest, nearly the whole of which is as well or better fitted for settlement than the best parts of Ontario, Nova Scotia or New Brunswick, and is ready for the plough.

We doubt, indeed, if the fertile area of the Canadian Northwest, climatically adapted for settlement, is as large as one of the small States of the American Union. The great elevation of the country, and its high northern latitude (hence the severity of the climate), to say nothing of the grasshopper plague and the scarcity of wood and water, compel us to dissent from Mr. Bennet's figures.

With regard to the aggregate area of the settled provinces, he said: "from a survey of the whole we may conclude that not less than a sixth part of it is capable of supporting an agricultural population, and that in the next century we may have in it a population of forty or fifty millions of people."

Mr. Bennet's calculations are marvellous; let us forget them, and take the following appropriate advice. He said: "We must see that we have a cultivable territory, as the hardiest races will find it impossible to compel granite rocks, sandy plains, and ice bound soils to yield sustenance to man."

In his work, "Ocean to Ocean," the Rev. G. M. Grant indulges in visions similar to the foregoing. In page 179 he says: "We have a great and fertile Northwest, a thousand miles long and from one to four hundred miles broad, capable of containing a population of millions." That is, about one hundred and sixty millions of acres of fertile lands.

At a public meeting held in Inverary, in Scotland, in 1876, the Rev. Dr. Taylor, an Emigration Agent for Canada, is represented by the *Oban Times* to say that "more than 111,000,000 acres of excellent land had lately been thrown into the market by a treaty with the Indians." He said he had seen in the Canadian Northwest "rolling uplands of enormous sweep, enclosing at one range of vision, and at more than one point, a basin one thousand square yards in extent, every yard of which could be ploughed. . . . In one day he had seen herds of from 80,000 to 100,000 bison (buffaloes) roaming about over its undulating plains, while millions of prairie-birds and waterfowl were met with at intervals all over the territory." Comment is unnecessary.

At the close the Marquis of Lorne, who presided at the meeting, very justly recommended that those in search of homes "should always make sure of knowing as much as possible about the country they purpose to go to. They should ask all about it from their friends who had already, perhaps, gone out, and that would help to prevent disappointment."

In one of the numerous pamphlets recently published relating to the Northwest, entitled "Manitoba and the Canadian Northwest," 1877, p. 15, it is said "that the total area of the lands known to be fit for cultivation is estimated at 375,184,000 acres." And in the Peace River region it is asserted that there are "thousands of acres of crystallized salt," and "thousands of acres of coal oil fields were found, the tar lying on the ground being ankle keep." Another of these equally unreliable authorities, entitled a "Guide to Manitoba and the Northwest," gives as Professor Macoun's opinion, "that the farther one went north the warmer the summers became." In a pamphlet containing a series of "Letters by James Trow, M. P.," and published by the Department of Agriculture for Canada, for 1878, p. 23, it is said that "the last census returns show that Canada produced one-sixth as much wheat, one-fourth as much oats, one-third as much barley, and as many peas as the thirty-four States and seven Territories of the United States." The forty-seven States and seven Territories of the Union in 1870 produced the enormous number of 287,745,626 bushels of wheat, 282,107,157 bushels of oats. In 1871 Canada, exclusive of Manitoba, whose product is not given in the census, raised 16,993,365 bushels of wheat, being about one-sixteenth that of the United States—45,610,029 bushels of oats.

But few, however, have exceeded Lord Dufferin in this kind of patriotism. His descriptions of Arctic scenery are very lively. His "Letters from High Latitudes" show, says the London *Times*, "that he can be cheerful amid the horrors of a Spitzbergen winter, and he thought Iceland a charming place."

Lord Dufferin visited British Columbia in the summer of 1876, and the Red River settlement in the following summer. Though it is well known that the Pacific Province is a "sea of mountains," and almost worthless as an agricultural country, and that the section of the Canadian Pacific railroad in that Province will cost the Dominion about three millions of dollars to every thousand inhabitants British Columbia contains, or one million for every thousand it will, in all probability, contain half a century hence, his Lordship said: It "is a glorious Province, a Province which Canada should be proud to possess, and whose association with the Dominion she ought to regard as the crowning triumph of confederation."

If British Columbia is the crowning triumph of confederation, the old Provinces of the Dominion pay well for it when they sink thirty-five millions of dollars in constructing a railroad among its mountains.

His speech in Winnipeg is, however, a gem in its way. His Lordship undertook to correct the "inadequate idea," which, he said, is "entertained by the best educated persons of the extent of Her Majesty's North American possessions," "by a summary of the rivers which flow through them."

The length of the river St. Lawrence and its chain of lakes to Thunder Bay, Lake Superior, he gave at 1,500 miles, and its main tributary, the Ottawa River, at 550 miles.† Of the water communication between Lake Superior and Red River, he enumerates the Kaminuistiqua River at 100 miles in length; Lake Shebandowan and Rainy Lake and River, 200 miles; he refers to the Lake of the Woods, and "from this lacustrian paradise of sylvan beauty" he passes "to the Winnipeg, a river whose existence in the very heart and centre of the continent is in itself," he says, "one of nature's most delightful miracles, so beautiful and varied are its rocky banks." Red River is given at 500 miles in length; the Assiniboine, 480; Lake Winnipeg, 300 miles long by 60 broad; the Saskatchewan, 1,500; he refers to the Arthabaska River; the Mackenzie River, 2,500, and the Fraser River is given at 1,400 miles in length.

"In this enumeration," he says, "those who are acquainted with the country are aware that, for the sake of brevity, I have omitted thousands of miles of other lakes and rivers which water various regions of the Northwest,....but the sketch I have given is more than sufficient for my purpose; and when it is further remembered that the most of these streams flow for their entire length through alluvial plains of the richest description; where year after year wheat can be raised without manure, or any sensible diminution in its yield, and when the soil everywhere presents the appearance of a highly cultivated suburban kitchen garden in England, enough has been said to display the agricultural riches of the terri-

11

tories I have referred to, and the capabilities they possess of affording happy and prosperous homes to millions of the human family."

This enumeration of the chief lines of water-courses shows undoubtedly that Canada is an immense Dominion, but to say "that the most of these streams flow for their entire length through alluvial plains of the richest description," to say that "the soil everywhere presents the appearance of a highly cultivated suburban kitchen garden in England," is certainly saying far too much. •

We have yet to learn where there is a river of any note in the Dominion which flows for even half its length through alluvial plains of the richest description. The St. Lawrence runs through an immense region of worthless country. The Ottawa and its numerous tributaries are in a Laurentian region. The streams and lakes between the Ottawa and Red River valley, a thousand miles, are also in a Laurentian region, containing but few spots fit for settlement. His Excellency omitted to say that the country adjoining Rainy Lake and River, and that adjoining the Lake of the Woods, on the south, is United States property. The Red River is the only stream of note within the habitable part of the arctic slope which flows through alluvial plains for the greater part of its length. But he omitted to say that four hundred miles of this river is in the United States, and that of its fertile valley, which is 315 miles in length, two hundred and fifteen miles are in the Republic. The Assiniboine and some of its tributaries flow through an area of good land, as well as sand plains and other infertile lands. The south branch of the Saskatchewan is chiefly in the great desert. The north arm of this river is the only stream in this part of the Dominion which passes through any great extent of fertile land. And this tract does not possess the agricultural capabilities ascribed to it. It is far north and is elevated from 1,500 to 3,000 feet above the sea, and contains large areas of infertile soil.

It is said there is a large extent of fertile land in the Peace River valley; and there may be tracts of fertile lands in the valleys of the Arthabaska, Mackenzie and other Arctic rivers of the far north, but the Lord help those whose lot may be cast for a winter in those icy regions. And as to the Fraser and other rivers of British Columbia, they are in the most worthless region, in an agricultural point of view, to be found on the American continent.

Such speeches as that of Lord Dufferin's at Winnipeg, however well intended, are a positive injury to a country like the Dominion. They only tend to cherish delusions, and often lead to disappointments of a serious nature. No country in the world has suffered more from such visionary views than British North America.

In its review of the Governor General's speech, the London *Times* said: "What Lord Dufferin tells us is that we possess" in North America "space and material capabilities for a population many times larger than our own, in a climate not more ungenial than that endured by a large part of these islands. We are proud of our rivers, our few lakes, our scenery, and our many triumphs over nature." He tells us, says the *Times*, that "the Canadian Dominion surpasses us in all these points. We have there all that we prize preëminently, that which we seem to prize most of all, whether individually or as a nation—illimitable space.... Art having done its work, nature is to resume its sway. That might seem impossible, but Lord Dufferin indicates a sense in which even this extravagant aspiration may be fulfilled, not here and there by a party of settlers, but by the whole population of the British Isles a dozen times over."

That is, the Dominion is capable of sustaining four hundred millions of people. Probably twelve or fifteen millions would be as many as the Dominion could sustain by means of its own resources.

The London *Times* continues: "We have all been wanting another England, Ireland and Scotland in the air, and there they are, safely landed within our own latitudes, a three weeks voyage off. The succession of enormous distances and strange surprises through which Lord Dufferin takes his hearers reads more like a voyage to a newly discovered satellite than one to a region hitherto regarded simply as the fag end of America, and a waste bit of the world."

Lord Dufferin has recently finished his term as Governor General of British North America. His presidency has been marked by great statesmanship; and as a citizen of the country, which he truly was, his lofty eloquence, lively conceptions, and brilliant imagination, were fully employed in favor of the Dominion of Canada on every befitting occasion; consequently he inherited the best feelings and highest regard of all classes of the Canadian people.

# CANADIAN PACIFIC RAILROAD.

During the three years following 1857, Captain Palliser was employed by the British Government in exploring the country between Ottawa and the Pacific Ocean, with a view of finding a practicable route for a railroad between these two points. After making a careful survey of this vast region, he declined to "recommend the Imperial Government to construct, or it may be said, to favor a thoroughfare by this line of route, either by land or water".... "Nor can I," he said, "advise such heavy expenditure as would necessarily attend the construction of any exclusively British line of road between Canada and the Red River settlement." He says: "My knowledge of the country, on the whole, would never lead me to advocate a line of communication from Canada across the continent to the Pacific exclusively through British territory. The time has now forever gone by for effecting such an object; and the unfortunate choice of an astronomical boundary line has completely isolated the central American possessions of Great Britain from Canada in the east, and also almost debarred them from any eligible access from the Pacific coast on the west."

However, strange to say, in less than ten years after Captain Palliser made his report, and before making additional surveys, the Dominion Parliament passed Acts providing for the construction of a railroad through this region, besides branch lines, and the whole to be completed in ten years. In consequence of the great difficulty of finding a suitable route, more than five years expired before the work of construction was commenced.

Such is the character of the country that after seven years spent by a large number of surveying parties, at a cost of $3,410,895 up to July 1877, the line had not been located up to the same month in 1878. The cost of the survey alone will probably amount to four millions, a sum sufficient to construct one hundred miles of railroad in the Lower Provinces of the Dominion. If the amount of time and money spent in making the surveys is a fair index to the future, the cost of constructing the Canadian Pacific railroad will be enormous.

The length of an air line between Ottawa and Burrard Inlet, on the east side of the Strait of Georgia, is 2,160 miles; but the shortest route surveyed between these two points is 2,774 miles. The variations in altitude and the amount of curvature of the line, especially in the mountain regions, together with the great extent of tunnelling, point to immense cost of construction and maintenance. Add to this the Red River branch, 81 miles, and the Lake Superior branch, both under construction, and we have 2,890 miles of railroad to construct between Ottawa and Burrard Inlet at the Strait of Georgia. And still the western terminus by railroad route is 234 miles, besides thirteen miles by navigation, east of Esquimault, the nearest suitable harbor on that part of the Pacific coast, making a total of 3,124 miles of railroad, exclusive of bridging the Strait of Georgia.

Up to February 1877, the date of the Chief Engineer's last report, all the lines through British Columbia unite in the Yellow Head Pass, in the Rocky Mountains, and vary in length, in the mountain region, from 464 to 550 miles. However, in consequence of the great difficulties of an engineering nature, and the proximity of the most favorable route to the United States boundary, every effort has been made to find a more northern route. The terminus at Burrard Inlet is close to the international boundary for a considerable distance at the Pacific end of the line, and against that at Bute Inlet there are serious objections. The cost of the work in British Columbia alone has been estimated at $36,500,000 to Burrard Inlet, or $34,400,000 to Bute Inlet. These are enormous figures, considering the distance. The terminus at Burrard Inlet has been adopted as the only feasible one at the Pacific end of the line. It is now estimated that the remaining part of the line to Esquimault would cost about twenty-seven millions of dollars.

It is obvious that the difficulties of obtaining a practicable route through British Columbia, and an ocean terminus on the Pacific coast, are unprecedented. The Strait of Georgia, though chiefly in the Dominion, is almost useless as a path for ships between the main land and the ocean. The British naval authorities on that station say that the navigation of the Straits by the north of Vancouver Island "is decidedly unfavorable," and the approach by the South of Vancouver Island to the Strait of Georgia is through passages more or less intricate. Still, the main passage is capable of accommodating large class vessels. But the passages are between, or at no great distance from the San Juan Islands, which belong to the United States. "All the naval authorities admit," says Mr. Fleming, "that vessels on their course" to the inner terminus of the Canadian Pacific railroad "would be exposed to the guns of the United States in the event of hostilities." And Rear Admiral A. De Horsey, who made a survey of the approaches in 1877, rejects "all idea of a terminus" for the railroad on the north of Vancouver Island, and says of the south passage that the San Juan and Stuart Islands "form the key of the navigation inside Vancouver Island. In the case of war with the United States that power might readily stop our trade through Haro Strait [San Juan was visited last month, September, 1877, by General Sherman, I believe with a view to its fortification.] The possession of San Juan might enable the United States, in case of war, to cut off our supply from the coal mines by sea." He therefore recommends the adoption of an outer terminus for the railroad as the only convenient one. But this involves thirteen miles of navigation across the Strait of Georgia, and the construction of 234 miles of additional railroad, 183 of which is on Vancouver Island, making the total length of the Pacific railroad and branches 3,124 miles. It has been proposed to bridge Seymour Narrows, a distance, says Admiral Horsey, "of 2,575 feet, in two spans of respectively 1,200 and 1,350 feet. To execute this work the middle pier has to be erected on a rock said to be eighteen feet under water at low tide, with a velocity of tide over it of from five to eight knots. This," he says, "would be a work of vast magnitude and expense, even if it be practicable to place a foundation on the rock; which I doubt, as there is hardly any slack tide. Nor must it be forgotten that bridging Seymour Narrows would, as regards large ships, obstruct the only practicable channel between Vancouver Island and the main land. This alone should, in my opinion, preclude its attempt." Mr. Fleming says: "The bridging from the main shore to Vancouver would be unprecedented in magnitude, and that its cost would be indeed enormous."

Thus, the Pacific end of the Canadian railroad will not have an ocean terminus, unless a bridge shall be constructed across the Narrows, and the railroad extended to the ocean. In place of a bridge, Admiral Horsey recommends the construction of a pile dock terminus on each side of the Strait of Georgia, and "ferrying a train" from one terminus to the other, a distance, he says, of thirteen miles, being four miles wider than the winter mail route between Prince Edward Island and New Brunswick. Some such means of transit as that named by the Admiral may possibly be required towards the end of the next century, but during the remaining part of the present century, at least, the resources of the Dominion will be taxed to their utmost in order to construct the section between Lake Superior and the Strait of Georgia, a length, together with the Red River branch, of 1,060 miles.

However, as Mr. Fleming says, as it has "been determined to establish a railway through Canada to the Pacific coast," it is important that we should have some idea of the burdens to be imposed on the tax-payers of the Dominion. The cost has been variously estimated at from $140,000,000 to $160,000,000, and as high as two hundred millions. Indeed, such is the nature of the work it is impossible to estimate the cost at present. Assuming that the Canadian Pacific railroad and branches can be constructed for the same price per mile as the Intercolonial railroad, say $44,000, one hundred and forty millions ought to be sufficient, exclusive of the cost of the Seymour Narrows ferry. But when we consider that the Intercolonial railroad is in a comparatively level country, thickly settled for considerable distances along the line, and that no part of it is far from the base of supplies, and on the other hand, looking at the remote situation of the Pacific route, the paucity of population, and the almost insurmountable obstacles of an engineering character in the way, this estimate is too low by a sum not less than from $18,000 to $20,000 per mile, in British Columbia alone; and when we consider that in this region the Coast and Sierra Nevada ranges of the South are united in one sea of Cascade mountains, besides the Rocky and numerous subsidiary ranges, Mr. Fleming's estimates may be too low. One line, 493 miles in length, he estimates at $35,000,000; another, 464 miles, at $39,000,000, and a third, 550 miles, is estimated to cost $28,000,000. The cost of tunnelling in the mountain regions through which the Pacific line is run will be immense; and the cost of bridging will have but few parallels. On the plains alone, the aggregate length of the bridging is over two

and a half miles, and the height of the bridges varies from
forty to one hundred and ten feet. The parts under construction are comparatively cheap lines to build. The bridges are
being constructed of wood, which in consequence of the
scarcity of suitable timber along the line on the plains, cannot
fail to be costly. Where suitable timber is plenty and convenient, bridges constructed of wood only cost half the price
of iron bridges, but, according to Mr. Fleming's view of such
works on the Intercolonial railroad, it is "a very grave error
to build the bridges of wood." "The saving effected in the
first outlay from the employment of wooden bridges would be
very small, not to be mentioned as an equivalent for the reduced standard of the railway, or for the increased annual
charge for repairs and renewals, for the risk of accidents by
fire, or for the increased danger of life and property." He
says: "The average length of life of a wooden bridge has
proved to be about twelve years." And that "it is apparent
that one-twelfth, or eight and one-third per cent. of the whole
cost is chargeable against it every year for renewals." Probably
there is no part of the world where the danger of destruction of
bridges by fire is greater than on these plains. Fires, once
started, have an unbounded sweep over the vast plains of the
west and northwest.

And the wilderness regions, east and west of the plains, are
largely covered with forest growth. That part between the
plains and Ottawa is generally described as a boiled-up sea of
crystalline rocks. Between the rocky ranges there are numerous
and extensive swamps, and the rocky ridges tower in height
far above the swamps and the deep valleys, which in many
places intervene. Some of the ranges lie "at nearly right
angles to the general course of the railway line, causing great
variations of altitude, with occasional high gradients, involv-
ing a considerable proportion of heavy work," including
tunnelling.

In the absence of an official estimate of the cost of this railway, one hundred and sixty millions of dollars may not be too
high. We may here observe that the greater part of the public works in the Provinces have cost more than the original
estimate. The Intercolonial railroad, for example, according
to Major Robinson's report, should have been constructed,
from Halifax to Quebec, for about $24,000,000. It cost,
including the section constructed by the Grand Trunk

Company, and taking into account the additional distance built, forty-seven mi' 3 at least, four millions more than Major Robinson's estimate.;

The Premier of the Dominion, Mr. Mackenzie, in a public speech in Sarnia in 1875, said: "We shall be able to accomplish the connection of our western with our eastern waters by 1890." That is 2,000 miles of railroad from Lake Superior to Burrard Inlet, including the Red River branch, is to be completed in about fift  years. The Intercolonial railroad, between Riviere du Loup and Truro, which is not a quarter of this distance, occupied seven years in building. Thus, one hundred millions of dollars, exclusive of the cost of enlarging the St. Lawrence canals, is to be added to the debt of this country by 1890. And when built, this long section will be almost useless, except to the United States, during half the year, when the navigation of Lake Superior is closed. The cost of working the road from Lake Superior to the Pacific waters, and the interest on the capital, will be a very heavy charge indeed. If the cost of working the whole road from Ottawa to the Pacific will amount to "not less than $8,000,000 per annum," and the interest on the capital to $8,000,000 more, as stated by the Premier, the Dominion will gain most when her Pacific railroad is not in operation. Against these large sums two millions is set down for revenue.

Evidently the inhabitants of the Dominion, especially those of the Atlantic Provinces, have not begun to count the cost of constructing and maintaining this gigantic work, nor have they generally estimated the end to be accomplished by its construction. When we consider that the main line passes, for one thousand five hundred miles, through mountain regions of the most worthless character, and that hundreds of miles of the remaining distance are in swamps and other infertile regions, and looking at the northern aspect of the whole country, the great depth of snow which falls in the mountain regions, it must be obvious to every thoughtful mind, especially to the tax-payers of the Dominion, that the Canadian Pacific railroad will be the greatest mark of human folly, except the tower of Babel, ever constructed. If there was any prospect of trade at the terminus, or at any intermediate point on the line, some hope might be entertained that in the distant future this road might pay half its working expenses. But when we consider that it begins nowhere and ends nowhere, so far as

12

regards future population or trade, it is vain to hope. British Columbia, containing in 1878 only about twelve thousand inhabitants, never can have more than a few isolated settlements here and there among its mountains. And these have the United States settlements on the Pacific coast as their immediate neighbors, and it never can be otherwise. The nearest settlements in Canada will always be remote from each other, and the road through the mountain region to them will be precarious. And the natural line of communication between Manitoba and the other communities of this continent is not with Canada in the east, but through the United States on the south, by the railroads now being constructed in that direction. The fertile part of the Red River valley is more than two thirds of its length in the United States, hence the adjoining State of Minnesota is the nearest neighbor to the Northwest settlements of the Dominion. Canada on the Pacific, and Ottawa in the east, and the Red River region in the centre, are too far apart to be neighbors. By the close of the present century a railroad may be constructed between the Straits of Georgia and Lake Superior, by means of which and the navigation of Lakes Huron and Superior for a distance of about 605 miles, communication with the eastern Provinces would be open for half the year. During the other half all communication would be through the United States. When we compare the Canadian Pacific route with the Southern, Central and Northern Pacific railroads of the United States, we may be told that the three latter lines also pass over extensive regions of desert and rocky country. True; but all these lines connect with eastern railroads passing through the most extensive regions of fertile lands to be found on the American Continent, inhabited by forty-five millions of people. The Southern Pacific road runs through an immense area of fertile country, and the Central Pacific. terminating in California, has a great future before it. This State is one of the richest, and is destined to be one of the most populous States in the Union. And the Northern Pacific railroad is constructed from Duluth, at the head of Lake Superior, 500 miles, to the Yellowstone River at Bismarck, 200 miles west of Red River. Upwards of twenty steamers navigate the Yellowstone to Bismarck. The Northern Pacific is connected at Breckenridge by a branch railroad from St. Paul, 216 miles in length, and also with the great network of railroads in the Western and Eastern States. From the 100th degree of longitude the Northern Pacific will pass through an immense region of infertile country, until it reaches the fertile valley of the Columbia River. The section between Bismarck and the navigable waters of the Columbia River will be constructed in a few years. The distance between the mouth of this river and Esquimault, the future terminus of the Canadian Pacific railroad, is only 150 miles. Hence there will be two competing lines of railroad, near together, for the Asiatic trade, both at about equal distance, 4,000 nautical miles, from Yokohama in Japan, or 5,350 from Shanghai in China.

The Northern Pacific railroad will be completed in all probability long before the Canadian line. The distance to build is not half so great as that of the Canadian, and the resources of the Columbia River region are much greater than those of British Columbia, and therefore afford greater inducements to construct the United States line. Besides, the comparatively broad valley of the Columbia affords an easy pass through the Cascade mountains. Already the United States have a railroad in operation from the Columbia river, about one hundred miles northward, with a view to its extension to the Strait of Georgia at British Columbia, thus preparing the way to divert the trade of British Columbia into the United States, as will also be the case with the trade of the Canadian Northwest. Flourishing towns and extensive settlements have already sprung up, both north and south of the Columbia River. This river and its branches are navigable for more than a thousand miles in the United States, besides for two hundred miles or more near its source, in British Columbia; thus opening up a passage for the timber, coal and other mineral products of the interior, to the ocean. Its waters teem with fish.

It is evident however, that two lines of railroad to the Pacific, so near to each other, and through so much worthless country, are not required. One line will be sufficient for fifty years to come.

If the Dominion desired to connect Manitoba by railroad with the railroads of the east, a much more feasible route might be found than the one proposed. By the end of 1878 the Red River branches, one on each side of the international boundary, will probably be finished, and connection made with Duluth, at the head of Lake Superior, a distance of about

455 miles from Fort Garry. And, "assuming," says Mr. Fleming, "the Duluth Railway to be extended along the south shore of Lake Superior to Sault Ste. Marie, the outlet of Lake Superior bridged at that point, and the railway continued thence into Canada by the north shore of Lake Huron, thus forming the most direct possible connection between Duluth and the cities of Canada." We add: a continuous line of railroad to Ottawa might now be in operation, the total distance being about 1,220 miles, one third of which is between Ottawa and the outlet of Lake Superior. Thus about 410 miles would be Canada's portion in the east. If this route had been adopted the Lake Superior section would be constructed by the United States, and the Dominion would save upwards of thirty millions of dollars in construction in this region alone, besides years and millions of dollars wasted in making surveys. But the Dominion powers have decided otherwise, and the people must submit.

When properly understood it is doubtful if a single electoral district could be found in the Atlantic Provinces that would favor the construction of the Canadian Pacific railroad. What benefit, direct or indirect, Nova Scotia, New Brunswick and Prince Edward Island can possibly derive from this work we cannot divine. One thing is clear, however, that being large consumers of foreign products, these three Provinces will have to pay a large share of the heavy taxes which future Parliaments will be compelled to impose. And it is doubtful if there are many constituencies in Quebec or Ontario which would vote in favor of its construction. The inhabitants in the valley of the St. Lawrence may derive some benefit from the expenditure of vast sums of money in the granite wilderness in their rear. But how long the people who do the work, and the money they may receive, will remain in the Dominion after the road is constructed, is a question easily answered. In a word, the Dominion is about to tax its people, and expend an immense sum of money for the benefit of the United States. But there has been a cry raised of loyalty to Canada, and for this the leaders of the two political parties in the Dominion have pledged the country to an expenditure for a worthless work equal to a war debt. As to the liabilities of the Dominion the reader is referred to the article, *Debt* and *Population*, in another part of this work.

## DEBT AND POPULATION.

"The chapter of National debts," said the Right Hon. W. E.
Gladstone, "is assuming....a painful and baleful prominence
as a social and political fact of modern experience." What he
said of "borrowing becoming the standing vice of almost every
country in Europe," is equally true in regard to the United
States and the Dominion of Canada. Nowhere has "this
mischievous and injurious process" had so injurious effect as
in these two countries since 1860.
After showing the amount of national debt Great Britain
and some other European nations owed, he referred to the
debt of the United States as "something wonderful—wonderful as the creation of four years,....amounting to nearly
3,000,000,000 dollars. Looking at these figures, a man would
be struck with something like despair, but if we look at the
position of the country which has to bear the burden I must
confess that I think the future of America, as far as finance is
concerned,....will not be attended with any embarrassments.
I do not believe that that debt will constitute any difficulty
for the American people." He said: "I do not hesitate to
declare I contemplate with the least anxiety....the debt of the
United States." Such are the vast resources of the country
"that in a moderate time it will be brought within very small
limits, and may, even within the lifetime of persons now living, be effaced altogether." He said the United States raised
"the largest sum of any country in the world for the purposes of a central government."
Since Mr. Gladstone uttered these words the United States
has paid nearly five hundred millions of her national debt,
and has reduced her taxes largely, and thereby the cost of
living in the Union.
No countries in Europe have suffered more from war debts
than Britain and France, and none possess greater ability to
pay. The recuperative powers of these two nations have no
parallel in Europe. The speedy manner in which France
paid her immense war debt which arose out of her recent war
with Prussia, astonished the world. And Great Britain's debt,
which is about $3,900,000,000, though immense, is not looked
upon as too great a burden to the nation.

But the natural resources of both these countries together are limited when compared with those of the United States. The population of the latter already exceeds the aggregate population of the British Isles and the Dominion of Canada by eight millions, and may within the lifetime of persons now living exceed the total population of both Great Britain and France.

To what extent the burdens imposed upon a country by means of a large debt retard its progress is not easy to estimate. Where the resources are immense, as in Great Britain, France and the United States, the burden is comparatively light.

The United States tables, showing the amount of the public debt of that country from 1790 to the present time, are highly suggestive. The population of the Union at that date was about equal to that of the Dominion of Canada in 1878. The national debt of the former was then only $75,463,476, being less than half that of the Dominion at the latter date. At the beginning of the present century it was less than eighty-one millions. It rose to one hundred and twenty-seven millions of dollars at the close of the war with Great Britain in 1812–14. It then fell gradually, until in 1836 it was only thirty-seven thousand five hundred dollars. From this date it rose gradually to $64,842,287 in 1860, when the rebellion commenced.

It was during the time when she had comparatively no debt, no taxes, that the United States made her greatest progress. It was before she felt the burdens of a public debt that her population increased most rapidly. In consequence of the rebellion, her debt rose to $2,773,236,173 in 1866. On the first of March 1869 it was $2,525,463,260, and on January first, 1878, it was reduced to $2,045,955,442; thus there was nearly four hundred and eighty millions paid in less than nine years. In the years previous to the current depression in trade she paid nearly one hundred millions of dollars a year. The payments in the last year show signs of returning prosperity. If prosperous times immediately return, and continue, the United States debt ought to be reduced to a small figure by 1890, and paid in full before the end of the present century.

From this sketch of the debt of the United States let us see how the Dominion stands. While in the Provinces, we have been viewing the United States debt as enormous, it is doubtful if we are generally aware to what magnitude the public debt of Canada has been increasing.

At the formation of the Dominion in 1867, the public debt amounted to $78,688,756, ~~$78,688,756~~ $ 75.718.16/

Manitoba became a Province of Canada in 1870; British Columbia in the following year, and Prince Edward Island in 1873, consequently a large amount was added to the public debt. On the first of July, 1877, the net debt of the Dominion amounted to over one hundred and thirty-three millions of dollars, and the annual interest, which is nearly all paid in Europe, is about seven millions. To the present debt may be added thirty millions in consequence of existing contracts for enlargement of the canals of the St. Lawrence, and the construction of the Pacific railroad, which will make the net debt of the Dominion upwards of one hundred and sixty-three millions of dollars. This is a greater debt by thirty-six millions of dollars than the United States owed at any one time previous to the recent rebellion in the Union. Hence the public debt of Canada is nearly equal per head of population to that of the United States, without comparatively any resources to fall back upon as means of payment. At the present rate of increase of the Canadian debt, and decrease of that of the United States, the Dominion will, in a very few years, owe more than double, per head of population, than the United States.

If the Dominion fulfils her obligations to British Columbia, if she continues to construct the Pacific railroad, her chapter of public debt has only began.

In 1875 Sir A. T. Galt, one of the ablest and most far-seeing statesmen in the Dominion, said: "Respecting the liabilities of the country, I look with the greatest alarm at their rapid and enormous increase. Commenced by Sir John, and continued and endorsed by Mr. Mackenzie, they are augmenting in a ratio far exceeding any possible growth of our population or resources, and must inevitably soon reach such a point as will grievously press upon our industry. Though expenditure may for the moment add to the business activity of the country, and be useful at a period of serious commercial depression, yet if such outlay be not reproductive at a very early day it is evident that the taxation incident to it will prove an intolerable burden....I consider the proposition perfectly monstrous that for the sake of the sparse population on the Pacific coast, ~~that~~ the prosperity of the four millions of people east of Lake Superior should be arrested, and their political independence jeopardized....I believe nine-tenths of the people of Canada are convinced that the construction of the Pacific railway is, at this time, and will be for many long years, wholly unnecessary....The frank and honest course is to tell British Columbia that the engagement was improvident, and its fulfilment impossible; to offer reasonable equi. valents for its abandonment....It is, however, certain that even were the other engagements made and pledges given respecting the canal system, that and other works will task all the resources of the country for years to come."

Since Sir A. T. Galt penned these lines over seventeen millions of dollars have been added to the debt of the Dominion, besides what is due on existing contracts, while in the same time the population has increased but slowly.

The increase in the public taxes of the Dominion will be more readily understood from the fact that in 1867 the total outlay by the Central Government was about thirteen and a half millions of dollars, while last year it amounted to nearly twenty-three and a quarter millions. And in the face of this heavy amount of taxation, and the vast accumulation of debt, the volume of trade has within a few years decreased about fifty millions, and consequently there is a deficit of \$3,361,000 in the revenue in the last two fiscal years. Notwithstanding, the Dominion is under statutory obligations to construct about 2,890 miles of railroad between the Ottawa River and the Pacific Ocean. Thus the finances and obligations of this country have assumed a shape calculated to create alarm among its inhabitants as well as among the public creditors.

The cost of the Pacific railroad and branches is variously estimated at from one hundred and forty to one hundred and sixty millions of dollars. Whatever the cost may be the money has to be borrowed in Great Britain, and the interest, together with that on the present debt, has to be paid to European bankers.

Every budget speech by the Finance Minister shows how difficult and how costly an affair it is for Canada to borrow money. The Dominion has no wealthy bankers, like Australia, able to float her bills in the money markets of England, hence she cannot borrow on as favorable terms as Australia. Besides, the latter is far removed from the dangers of war, while an angry word on the part of the United States against Canada would close the money market of Britain against the

Dominion. And what would tend to increase the difficulty of borrowing is the fact, says the Finance Minister, "that $150,000,000 of British capital was invested in private enterprise in Canada, and which were paying no dividends, and bearing in mind the burdens we had to bear, it was not to be expected we could successfully compete with the Australians in the London markets."

Thus one of the most discouraging features in connection with the debt of the Dominion is that it has been chiefly incurred by means of the construction of works which have not paid the cost of management and repairs. Such is the case with regard to the canals of Canada. The Intercolonial railroad, and the Prince Edward Island railroad, British Columbia, and the Northwest, continue to cost immense sums over and above income.

The cost of working the Intercolonial railroad between Rivière du Loup and Halifax in 1877, was........$1,461,673
Revenue derived................................. 1,154,445

Expenditure over income..........................$307,228

And as to the Pacific railroad there can only be one opinion, and that is, it will be absolutely ruinous to the Dominion.

The question is, who are to pay this vast amount of debt, present and prospective? Looking at the great increase in the human family within the last fifty years, and taking into account the large emigration from Europe in that time, we may fairly conclude that in the future the emigration from Europe and Asia will be much larger than in the past. In the British Islands, and in many other parts of the old nations, nearly all the available lands are occupied, hence profitable employment is not increasing in proportion to population.

The great facilities for travelling and transit in the present age will give an impetus to the migratory movement of the future. Countries which were remote even a quarter of a century ago, are now, by means of steam power, brought very near to each other; and by telegraph far distant countries hold momentary converse. Thus, by these and other means, Europe and America are united, and Asia is not far off.

Within the last sixty years nine millions of Europeans have settled in the United States and adjoining Provinces. Of these, eight millions went to the States. The influx has been gradual, and recently from almost all countries. A large proportion was from the British Isles, many came from Germany, and now distant China is sending a quota to California.

The immigration of Europeans has had a highly beneficial influence in keeping the progressive forces of civilization in North America rightly adjusted. All classes and nationalities have mingled and joined in the national life of the country of their adoption.

That the tide of human life will continue to flow from the shores of the old States of the world to the American continent, and at an increased rate, there is but little doubt. Europe is nearly full of people, and opposite the west coast of America there are more than six hundred millions of hungry Asiatics, millions of whom may find their way to America.

There are, however, many things connected with Asiatic life that may be obstacles in the way of a large emigration from that country at present.

Emigration is one of the great social problems of the age, and will continue to be so as long as there are fertile lands in America for sale. And the redundant population of the old States of the world will be compelled to emigrate. It is only the wretchedness and poverty of the people that prevents millions of them from leaving their homes at present. However, people desirous of fleeing from heavy taxes at home, may feel reluctant to emigrate to a country equally burdened by taxation and debt. It is therefore important that the taxation of these American nationalities should be kept within reasonable limits.

The following tabular statement shows the past, and an estimate of the future, population and debt of the Dominion of Canada for the years named therein:—

| YEAR. | POPULATION. | NET DEBT. | DEBT INCLUDING ASSETS. |
|---|---|---|---|
| 1861.... | 3,172,404 | | |
| 1867.... | ................ | $ 75,728,641 | $ 93,046,051 |
| 1871.... | 3,584,238 | 77,706,517 | 115,492,682 |
| 1872.... | ................ | 82,187,072 | 122,400,179 |
| 1873.... | ................ | 99,848,461 | 129,743,432 |
| 1874.... | ................ | 108,324,964 | 141,163,551 |
| 1875.... | ................ | 116,008,378 | 151,663,401 |
| 1876.... | ................ | 124,551,514 | 161,204,687 |
| 1877.... | ................ | 133,208,694 | 174,675,835 |
| PROSPECTIVE. | | | |
| 1881.... | 4,100,000 | $170,000,000 | ................ |
| 1891.... | 4,800,000 | 250,000,000 | ................ |
| 1901.... | 5,700,000 | 300,000,000 | ................ |

The population of the Provinces has been of slow growth during the last twenty years, since the fertile lands of Ontario have been disposed of; and it may be expected to be slow in the future in all the old Provinces, where there is but little fertile land remaining unsold.

And as to the Northwest it remains to be seen how far settlement will extend down the arctic slope. The obstacles in the way are very great indeed, much greater than in the eastern Provinces. The foregoing estimates of population for the present and two following decades are not half as large as those generally made by Canadian authorities, as will be seen by reference to other pages under the head of "Exaggerations."

In the foregoing table one hundred thousand may be added to each decade for Indians. As they do not pay taxes, but are themselves a heavy tax on the Dominion, we have not included them in the table.

If the Dominion is not progressing rapidly in population, she certainly is progressing in the accumulation of debt, as will appear by reference to the last table. The total debt, including what is due on existing contracts, is about two hundred millions of dollars. Against this amount there are assets as shown in the table; but some of them are not worth much as a means of payment.

As, according to the Chief Engineer's Report, it has "been determined to establish a railway through Canada to the Pacific coast," and that the part between Lake Superior and the Pacific is to be built by 1890, as announced by the late Premier, we have based our estimates of prospective debt on these enunciations. However, it would be well to extend the time when this section shall be finished to the end of the present century.

Evidently boasting and borrowing have become standing vices in the Dominion of Canada, and we sometimes boast most about that which is most worthless.

And it is now obvious that the population of the United States in 1880 will be more than ten millions less than that estimated by statisticians of that country. This is a great difference in ten years. Indeed, the population of the Republic may not exceed fifty-eight millions in 1890, or seventy millions at the end of the present century; that is upwards of thirty millions less than that estimated by United States

authorities. Sir Charles W. Dilke, in his "Greater Britain,"
dated 1869, says: "At the present rate of increase, in sixty
years there will be two hundred and fifty millions of
Englishmen dwelling in the United States alone." This
estimate probably is too large by more than one hundred
millions.

The United States and the Dominion are so remarkably
united by nature, as well as by roads, railroads and other
means of communication, and also by commercial interests,
and the Provinces are so largely dependent on the States, that
it is almost impossible for the latter to develop their own
resources unless in a commercial union, at least, with the
United States. Without a union of interests, the Provinces
cannot retain their own population. Nothing less than a
market in the States as free as that between State and State
will meet the growing wants of the Dominion.

And such is the social character of these two nationalities,
that the conventional line that divides them is disregarded by
their inhabitants except in their commercial relations. There
are comparatively few families in the Provinces but what have
relatives in the United States. And the laws and institutions
of the two countries are so nearly alike that their inhabitants
pay but little respect to any distinctions that may exist. There
is a general commingling of the adjoining peoples; indeed, in
some places it is not easy to make out who belongs to the
States and who is Canadian. Some farms are divided by the
international boundary. The patriotic convictions of many
of the people along the boundary are a matter of mere convenience or self-interest. And of the immigrants, many enter
the States by way of the River St. Lawrence, and others pass
through the United States on their way to Canada. By these
and many other ways the States and Provinces are gradually
becoming one people.

And immigrants from the British Islands largely prefer the
States to the Provinces as a home; and those coming from
other countries to America nearly all go to the United States.
In 1870 there were 2,626,241 natives of the British Isles in the
United States. Of these 1,855,827 were from Ireland, while
in 1871 there were only 485,524 from these Isles in the four
original Provinces of the Dominion.

The population of the States and Provinces has been largely augmented by immigration, as the following figures show:—

The number of foreigners who settled in the United States in the
    ten years ending in 1840, was....................... 599,125
In the decade ending 1850.............................1,713,251
    "    "    "   1860.............................2,598,214
    "    "    "   1870.............................2,491,451
In the six years ending 1876.............................1,896,782

Those of foreign birth residing in the United States in 1870 numbered 5,567,229.

*Canadian Immigration.*—The number of immigrants reported as having settled in Ontario, Quebec, Nova Scotia and New Brunswick, were:—

In the decade ending in 1861.........................232,000
    "    "    1871..................... ....199,093
In the six years ending 1877.........................206,098

These last numbers include 7,923 Mennonites and Icelanders who settled in Manitoba. The expenditure for the last six years named in the last table amounted to $1,515,676.

The total number of persons in these four Provinces who were born in other countries, was:

In 1861..............................................677,967
In 1871..............................................583,822
                                       ————
Decrease in ten years............................. 94,145

Thus there were ninety-four thousand less persons in these Provinces in 1871, who were born in other countries, than there were in 1861, while the number of immigrants was only thirty-three thousand less.

According to the rate of increase in these four Provinces between 1851 and 1861, there should have been nearly four hundred and forty-eight thousand more people in them in 1871, than there were in 1861. Probably a large number of the immigrants who are reported as settlers afterwards removed to the United States, and were recorded there as immigrants from Europe, though many of them may have resided in the Provinces for years.

The United States census for 1860 shows a population of 249,970 British Americans in the Union. In 1870 the number was nearly double, being 493,464. These facts show that the annual emigration from the Dominion to the States is fully as large as that from Europe to the Dominion. And this is confirmed by the report of the Rev. P. E. Gendreau, who was sent in 1873 to make "enquiry into the emigration which" was then "going on from our country to various points in the American Union. The total of 800,000," he said, "may perhaps be admitted as representing the population which has emigrated from Canada to the United States, including their descendants born in the latter country."

In the six years ending 1876, 204,611 persons removed from the four original Provinces of the Dominion to the United States; Quebec sent the largest quota. Recently a large amount of money has been appropriated, and every effort made by that Province in order to secure the return of Canadians from the States. A few have returned. But the exodus from the Provinces is still large.

In 1871 there were 64,447 natives of the States residing in the Provinces.

Looking at the configuration of the Dominion, the limited areas of fertile lands climatically adapted for settlement, and taking into account the present trade relations between the States and Canada, emigration from the Provinces to the States will be large, especially during periods of great agricultural and commercial prosperity in the latter country. And to give more force to this view of our future, the railway system of the old Provinces of Canada is nearly complete, and the lumbering business, except in remote places, is rapidly drawing to a close. Indeed, over large areas there is not more forest wood than is required for local use. The fisheries and mines afford but little employment in winter. Consequently employment at home for our young men, especially during our long winters, is becoming less plenty and less remunerative. And the financial embarrassments of the Local and General Governments, except that of Ontario, afford no hope of aid to more public works of any note for years to come. True, if some fortunate event does not intervene to arrest the construction of the Pacific railroad, a few out of the many may find employment on that work; but if its construction shall be continued, works that may be useful cannot be constructed by the Government for want of means, hence our young men will have to look to the States for employment. This is no imaginary picture; we are shut up to facts, and the sooner we know them the better.

Were it not for the large sums of money annually expended in promoting immigration, the decennial increase in the population would be very small indeed. How long the Dominion can stand the present and future strain remains to be seen.

After a careful consideration of all the facts at command, we are fully convinced that the Dominion does not possess a moiety of the resources claimed for it by those who are responsible for sinking the country so deeply in debt, and that it cannot redeem itself, unless it shall be allowed free and unrestricted access to the markets of the United States.

# TRADE AND COMMERCE.

There is no part of the world where a double line of Custom
House regulations is more vexatious than that between the
United States and the Dominion of Canada. And nowhere
on the face of the globe is there to be found a more unnatural
boundary line than that which divides these two countries.
A glance at a map of the country shows that the Atlantic
between New York and New Brunswick is the ocean front of
Ontario and Quebec, and that the United States railroads, and
the River St. Lawrence during half the year, are the only convenient lines of communication between these two Provinces
and the ocean. The States in front are their immediate
neighbors, and, in the event of free trade, would be their best
customers.

The lower or most eastern Provinces are all but disconnected
from the two upper Provinces. A long narrow range of almost worthless country, through which the Intercolonial
railroad passes, separates Nova Scotia and the chief settlements in New Brunswick from the settled portion of Quebec.
The lower Provinces have the States on the southwest as their
nearest neighbours and natural customers. These Provinces
are nearly as far from the Red River of the Northwest, and
nearly twice as far from Vancouver Island as America is from
Europe.

Looking at the isolated position of the habitable parts of the
Dominion, and at the nature and situation of its resources, there
cannot be a large intercolonial trade. Nearly all the forces
and resources of the country will turn naturally towards the
south—towards the United States. This broad opening for
commerce will clear the way for other and brighter results.
Nowhere will commerce prove to such an extent the necessity
for union as between these two nationalities. During half the
year the St. Lawrence is frozen, while the communication
through the United States will know no interruption. The
Canadian products for export must, to a very large extent,
enter the United States, either for consumption there, or on
the way to other countries. The chief part of the trade of

Ontario, Manitoba and British Columbia is with the United States, and must, even in the face of an almost prohibitory tariff, as at present, continue to be so. And if the lower Provinces make a grindstone, said the Hon. Joseph Howe, " it floats to the United States."

It seems, beyond any human possibility, that scattered communities, like those of the Dominion, could secure to themselves a free, prosperous, and undisturbed existence, unless in a commercial union with the United States. This is the only possible, because the only natural course. Trace as we may the boundaries of nations, we cannot find a country so peculiarly situated as Canada is. Indeed, the country is a mere geographical expression. To call it a nation is irony.

Though the Provinces have been benefitted in some ways by their political connection with Great Britain, it cannot be denied that they have suffered in many ways by that connection. Similar customs regulations to those which caused the old colonies to revolt were imposed by the Imperial Government on the trade of these Provinces. They were for a long time compelled to purchase nearly all their goods in the British Islands. Both Britain and the United States imposed high duties. Hence the trade between the Provinces and the States was but trifling. The States made rapid progress, while the Provinces remained almost stationary. All the Provinces got from the mother country in return was the protection of their timber and lumber in the British markets. But this advantage did not half meet the loss they sustained under Imperialt customs regulations.

In no way, however, have these colonies suffered so much by the connection as by the numerous treaties between the United States and Great Britain.

The treaties between these two nations form an important chapter in American history. Beginning with the treaty of peace in 1783; then followed the treaty of Ghent in 1814; the treaty of 1818, by which the fishery grounds were defined; that of 1831 referred the international boundary dispute to the King of Holland; in 1842 the celebrated Ashburton treaty was made, by which " British territory equal in size to two of the smaller States of the Union," was ceded to the United States; the treaty of Washington of 1846; the reciprocity treaty of 1854; and finally the Washington treaty of May 26, 1871.

14

The careful student of this part of American history can hardly fail to see that the United States obtained great advantages by means of these treaties. Even by the last treaty, in the making of which it was supposed the voice of Canada would have been heard, " territorial rights of great value," said the Canadian Government, were ceded to the United States. In this way the mother country obtained a settlement of her own disputes with the Union, which arose out of the Alabama and other troubles.

By the Washington treaty it is provided that the United States shall have the use of the Gulf and sea shore fisheries of Quebec and the Lower Provinces, except shell-fish, shad, salmon, and the river fisheries, for twelve years. Similar privileges are allowed to the inhabitants of the Provinces in United States waters. Fish oil and fish of all kinds, except fish of the inland lakes, and of the rivers falling into them, and fish preserved in oil, shall be admitted into each country free of duty.

It being asserted by the British Commissioners that the privileges thus accorded to the citizens of the United States are of greater value than those accorded to the people of the Provinces, it was agreed that the difference in value, if any, should be determined by arbitration. After the expiration of six years from the date of the treaty, an award of five and a half millions of dollars has been made in favor of the Provinces. But the United States protested against the award as being far too large an amount. It has been paid. And such is the conflicting nature of the international relations between these two countries that new difficulties will probably arise in relation to the fisheries which may lead to further trouble in relation to this vexatious subject.

In the settlement of the Alabama claims it was provided that a majority of the arbitrators were authorized to make an award; while under the article relating to the value of the fisheries no such provision is made for the payment of Canadian claims.

By Article 26 of the Washington Treaty, "The navigation of the river St. Lawrence, ascending and descending, from the forty-fifth parallel of north latitude, where it ceases to form the boundary between the two countries, from, to, and into the sea, shall forever remain free and open for the purposes of commerce to the citizens of the United States."

The navigation of Lake Michigan is free to the citizens of Canada for twelve years; and that of the other great lakes is open to the citizens of both countries, under certain conditions. Also the adjacent canals of both countries are free to the citizens of both; and also the transit of goods through both countries free of duty.' There are other but minor advantages allowed to the inhabitants of both countries.

It will be noted that while the river St. Lawrence is free, that of the Hudson from Albany to New York city is not free to the citizens of Canada. Hence, the navigation of the Erie Canal will be of but little benefit to Canadians.

By the treaty the export duty on United States lumber passing down the river St. John, amounting to about $65,000 per annum, ceased to be collected by New Brunswick.

It is now obvious, however, that the chief disputes as regards the fisheries remain unsettled. After 1883, when the present term of United States occupation of Canadian fishing grounds may terminate, all the old disputes about headlands and bays may be revived. The disputes over the fines to be paid on the renewal of the leases and licenses will, in all probability, again recur. Referring to this subject, the London *Times* said: " Once more we are confronted with the painful truth that the Treaty of Washington fails to provide for the settlement of the difficulty it was intended to remove. All the points in dispute are as much in dispute as ever." After a careful survey of these points, the *Times* said: " We confess we see no clear way out of these numerous difficulties."

The most important treaty, so far as the interests of British North America were concerned, was that of 1854, known as the "Reciprocity Treaty." During the operation of that treaty, when international commerce was allowed to follow its natural bent, at least in a measure, the domestic trade between these two countries rose from $20,691,360 in 1853, the year before the treaty, to $33,491,420 in the following year. In 1865. when the treaty was abrogated, the total imports from and exports to the Union amounted to $71,374,816; and in 1866, believing the treaty would be renewed, the trade between the two countries rose to $84,070,955. But in 1868, when all hope of a new treaty of reciprocity failed, the trade fell to $56,287,546, showing a decline of twenty-eight millions in two years. The total trade between these countries in the eleven years previous to 1855 was $163,593,435, while in the following eleven years, when the effects of the treaty were felt, their total trade rose to $554,631,910. And the ratio of increase was greater in the last than in the first years of the treaty. The limited compass and slow growth of their trade previous to 1854, its unprecedented expansion from that period to the close of 1866, when the effects of the treaty had reached their maximum, and the sudden decrease which followed the abrogation of the treaty, are significant results, and cannot fail to impress the inhabitants of both countries with the importance of closer free trade relations. And the trade between the Dominion of Canada and the United States did not exceed the value of $78,003,492 in 1876, and one million less than this sum in 1877.

Viewing the material progress made in both countries since the close of the treaty, we might safely conclude, had the treaty been continued in operation to the present time, that in place of seventy-seven millions the interchange of commodities between Canada and the Union would now be worth two hundred and thirty millions of dollars, or three times the total trade of the Dominion of Canada with all other countries in 1877. In 1854, the Provinces imported from the United States to the value of over fifteen and a half millions of dollars more than the value of their exports to that country. This was an unprofitable trade for the Provinces; but when the treaty became fully in operation the scale was changed, until in 1866, when their exports to the States amounted to $54,714,383, and their imports to only $29,356,572; thus a balance of over twenty-five millions was in favor of the Provinces.

With such results before us, the Government of Canada proper were justified in saying that "it would be impossible to express in figures, with any approach to accuracy, the extent to which the facilities of commercial intercourse created by the reciprocity treaty have contributed to the wealth and prosperity of the Provinces, and it would be difficult to exaggerate the importance which the people of Canada attach to the continued enjoyment of these facilities." The repeal of the treaty was "generally regarded by the people of Canada as a great calamity." And "if," said the Government, "the trade had been wholly unfettered and allowed to take its *natural* course," the results would have been much more favorable to the Provinces generally.

During the operation of the treaty the chief part of the trade between the Provinces and Great Britain became gradually transferred to the United States. And now, without the advantages which that treaty afforded, the trade between the Provinces and the States, even in the face of an almost prohibitory tariff, is comparatively large. The Dominion of Canada has used every argument and every influence in order to get the treaty renewed, but without success.

In an article relating to the Washington Treaty of 1871, the London *Times* said: "The ties between Downing Street and what used to be called 'our dependencies' are nominal." After referring to the effects of the treaty, said that "if the Canadian Ministry and Parliament had negotiated a Reciprocity Treaty with the United States—a necessary consequence of which would have been the adoption of a hostile tariff in Canada against England, for the purpose of developing freer commercial intercourse with the Federation—we should not have forbidden the treaty. In a word, it is competent to the Canadian statesmen to declare, and act upon the declaration, that their trading interests with the United States are more important to them than their trading interests with the United Kingdom." And Lord Kimberley, Colonial Secretary, fully realized the force of this view of our commercial relations when he said, "Her Majesty's Government are well aware that the arrangement which would have been most agreeable to Canada was the conclusion of a treaty similar to the reciprocity treaty of 1854." However the conclusion of such a treaty was a matter of small importance to Great Britain compared to the settlement of the Alabama and other disputes, hence his Lordship said: "Canada could not reasonably expect that this country should, for an indefinite period, incur the constant risk of serious misunderstanding with the United States, imperilling, perhaps, the peace of the whole empire, in order to endeavor to force the American Government to change its commercial policy." It is obvious, however, that Great Britain was willing to conclude a treaty of reciprocity for the benefit of Canada, even against her own interests at home, if it could have been reasonably effected.

Give Canada unfettered trade with the United States, and the patriotism of the Canadian people would soon rise above their loyalty to the British Throne. This Great Britain knows. Indeed, the only use Canada has for Great Briatain is what she can make by the political connection. It is not easy, at present, to define the commercial policy of the present Parliament of the Dominion; so far, however, as enunciated by a majority of the members, it is not Britain's free trade policy, but a system of protection, similar to that in operation in the United States, the Dominion is about to adopt.

We cannot point to a country whose trade relations are so much crippled as those of Canada. Not only in her. trade with the Union, but in regard to her forest exports. The most valuable forests of the Provinces are being rapidly destroyed in order to meet the pressing demands of their trade with Great Britain. The value of the exports of forest products to Britain has gradually risen in the last five years from thirteen and three-quarter millions to seventeen millions of dollars per annum in 1877. This might appear to be a profitable trade, especially when by means of it the exports from Canada to, and imports from, Great Britain are nearly balanced. Besides, it gives employment to thousands of men, and to a large part of the shipping of Canada. But the exports of forest products from the Provinces to Europe must soon be closed, and the sooner the better for Canada. Already Ontario is stripped of all her accessible timber. Nova Scotia and Prince Edward Island have none to spare, and the remaining forests of Quebec and New Brunswick are remote. The country to the source of their chief rivers has been stripped of nearly all the most valuable forest wood. And the class of wood products, deals and square timber, sent to Europe, has had a very destructive effect on the forests of this country, while the producers of lumber and owners of ships have been but poorly paid for their labour.

And during the same years in which the Canadian exports of forest products to Great Britain increased, her exports of wood to the United States decreased in value from thirteen millions to less than five millions of dollars in 1877. Thus, for the want of a free market in the United States, the forests of Canada are being wasted, and the lumber of the country is exported 3,000 miles to a market, where it does not realize more than half its value. And the lower the price falls in the markets of Britain the more lumber we export. As the forest resources of Canada is a subject of very great importance we shall advert to it in another page.

Taking into account the destruction of the forests, the trade of Canada with Great Britain costs the former more than it is worth. In 1877 Canada imported from the United States to the value of over twenty-five and a half millions of dollars more than she exported, and in the year previous to the value of fifteen millions.

And the trade between the Provinces themselves is but trifling. So far as known, there are only three coal areas in British North America; one near the Atlantic coast, another at the eastern base of the Rocky Mountains, and a third at the Pacific. The latter is about 800 miles from the Rocky Mountain coal field, and nearly 4,000 miles from the Atlantic coal areas. Hence the Province of Ontario, and, indeed, the greater part of the valley of the river St. Lawrence, and the Red River country are compelled to purchase their coal from the United States. The coal mines of Pennsylvania are near the Ontario settlements. Nova Scotia has abundance of excellent coal; and could easily supply the States fronting on the Atlantic, as far south as the city of New York, if it were not that three quarters of a dollar duty per ton has to be paid before a cargo of Canadian coal can be landed in the United States.

In 1877 the proprietors of coal mines in Nova Scotia petitioned Parliament to impose a duty on coal imported into the Dominion. The quantity imported from the United States in that year was 789,696 tons, including 415,869 tons of anthracite, and 20,032 tons of coke. The whole was valued at $3,176,154, while the exports of coal to the United States only amounted to 178,772 tons, worth $689,663. So far as known, there is no anthracite coal in the Lower Provinces; hence the importations of this class of coal into the Dominion is very large; even Nova Scotia and New Brunswick imported anthracite coal from the United States to the value of $137,383 in 1877. Of the importations Ontario imported coal from the United States amounting to 623,187 tons, valued at $2,506,244, and Quebec 126,067 tons, worth $510,559.

If a duty, equal to that imposed on foreign coal by the United States, shall be imposed on foreign coal entering the Dominion, the inhabitants of the St. Lawrence valley would have to pay over half a million dollars into the revenue of the Dominion, or import their coal from Nova Scotia at a greater cost.

But there is another side to the subject. Ontario has flour to sell, and the Lower Provinces of the Dominion are purchasers of breadstuffs from the United States to the value of $1,876,146. Hence, if Ontario has to pay a duty on coal, it would only be fair play to protect her breadstuffs by imposing a duty on importations from the States, which duty the Maritime Provinces would have to pay into the Canadian treasury.

In 1877 the Dominion imported wheat, flour, corn, and meal of all kinds, to the value of $13,200,384; of this the value of $12,958,945 was from the United States. Of this large amount the wheat imported was valued at $3,992,793, and Indian corn at $3,236,864. In the three years previous to 1878 the Dominion imported wheat, wheat flour, corn and corn meal, to the value of $36,018,727, or an average of twelve millions of dollars worth per annum; and exported of breadstuffs to the value of $34,322,404, or eleven and a half million dollars worth per annum. But a large part of the Canadian exports consisted of barley, oats, peas and beans, valued at $7,428,762. In this year Ontario exported of these and other coarse grains to the value of $5,429,774. Of barley alone, she exported 6,042,632 bushels; and imported of breadstuffs from the States to the value of $2,228,565 more than she exported.

The large importations of wheat and corn from the States, free of duty, afforded profitable employment to the Canadian millers and cheap bread for the Canadian people. A large part of the corn imported was used as feed for farm stock. Consequently Ontario and Quebec were enabled to increase their exports to Britain and other places in consequence of importations of breadstuffs from the United States.

Such are the requirements of the Dominion—such is the relative position of its habitable parts, and such are its conflicting interests, that a duty on coal would be a burden to Ontario, Manitoba, and a large part of Quebec; while a duty on breadstuffs would be equally burdensome to the Maritime Provinces, British Columbia and Quebec; and would deprive Ontario of the profits arising from her importations of wheat and corn.

The United States, with a view of promoting material industry, imposed a high scale of duties on such manufactures and productions of foreign countries as compete with like productions in the Union. How far this policy has been conducive to the interests of the Republic, is a question much more easily asked than answered. The Dominion has not yet adopted protection as it is understood in the United States; nor, with her comparatively limited resources, could she do so without injuring her trade. In a recent speech in Toronto, the late Premier of the Dominion, said, we should "keep our policy in harmony with that of the Mother Country in trade and in everything else where it is possible for us to act in unity with her."

There are but few things indeed, in which the Dominion can act in harmony with the British Isles. The simple fact is, Canada is building up the industries of the United States. Canadian imports from the latter country have kept gradually increasing, while her exports to the States has been falling off rapidly. In Canada, the United States have a free and convenient market for coal, grain, flour, and many other products; and it never can be otherwise. Canada cannot protect her own chief industries against the United States, unless at the expense of both her own and the British people.

The United States coal-fields are immense; much larger, it is said, than those of all the world put together. A well-informed writer says: "There are two hundred thousand square miles of coal lands in the country, ten times as much as in all the remaining world." Between the Atlantic and the Rocky Mountains, her coal fields are scattered here and there in all directions; and near the Pacific coast she has abundance of coal. And the means of transit by rivers, lakes, canals, and railroads, are so wide-spread, that both the States and the Dominion can be readily supplied. Ontario has to depend upon the Union for all her coal; and the Cleveland and Erie railroads have brought Montreal and other parts of Canada into near connection with the anthracite coal fields of the United States. And the agricultural products of the Union are so varied, and the quantities produced so immense, that any products of a like kind the Dominion may have to spare would add but little to the volume of the United States products; and consequently would make but little difference, if any, in the price of such products in the markets of the Union. In 1868, the United States exported agricultural products to the value of $319,000,000; and in 1878, to the value of $592,475,813; being nearly thirty times as much agricultural wealth as the Dominion exported in the same year, including imports from the United States.

15

Canada is in a very different position, in her trade relations, from those of either the United States or Great Britain. The latter has acquired a commercial prestige throughout the world. Before, and since she adopted free trade, Great Britain continued to manufacture largely for the world, and thereby accumulated immense wealth. The Canadian settlements are isolated, and confronted by the almost prohibitory barriers of the United States commercial system; so much so, that Canada's independent action, and even existence, as a separate nation, is almost, if not quite, impossible.

However, a system of trade which would work well in an old country like Great Britain might be a total failure in the United States, where forty-seven States and Territories, each the size of Spain, and containing an average of about one million inhabitants, have free and unfettered trade with each other. In no other part of the world is there such a magnificent trade area; such a shapely consolidation of States; such a variety of climate and productions, and such vast resources. When in Scotland, in 1875, the late Premier confirmed this view, when he said the United States "is a great country; its resources are enormous, its riches are almost incalculable." And in Sarnia, in the same year, he said: "We have a border of thousands of miles alongside the United States. The same boundary extends so far that it is impossible that any policy adopted by the United States Government can do otherwise than affect us either prejudicially or favorably."

Consequently, how the Dominion can keep her policy in harmony with that of the Mother Country in trade or in anything else we are unable to understand. It must be obvious to those who have carefully studied the resources of the Dominion, and the geographical position and relations of its habitable parts, that these provinces are not dependent upon each other, nor on Great Britain for profitable trade; but they are depending upon the United States. The chief settlements of the Dominion are too far scattered to be of any great benefit to each other in the way of trade. Mr. Richard Cobden took the correct view of this subject when he said: "That nature has decided that Canada and the United States must become one for all purposes of inter-communication. Whether they also shall be united in the same federal government must depend upon the two parties to the union." And Sir Charles W. Dilke, in his *Greater Britain*, referring to the trade between Great Britain and her colonies, says, p. 382,383, that " the retention of colonies at almost any cost has been defended—so far as it has been supported by argument at all—on the ground that the connection conduces to trade, to which argument it is sufficient to answer that no one has ever succeeded in shewing what effect upon trade the connection can have, and that as excellent examples to the contrary we have the fact that our trade with the Ionian Islands has greatly increased since their annexation to the kingdom of Greece, and a much more striking fact than even this—namely, that while the trade with England of the Canadian Confederation is only four-elevenths of its total external trade, or little more than one-third, the English trade of the United States was in 1860 (before the war), nearly two-thirds, and in 1865 (first year after the war), again four-sevenths of its total trade. Common institutions, common freedom, and common tongue have evidently far more to do with trade than union has; and for purposes of commerce and civilization America is a truer colony of Britain than is Canada."

" It would not be difficult, were it necessary, to multiply examples whereby to prove that trade with a country does not appear to be affected by union with it or separation from it."

It is now obvious that without much less restricted trade relations with the Union than those of the present, Canada cannot develop her own resources. Even in the event of an unfavorable grain crop in Ontario, as was the case in 1876, the other provinces would have to depend entirely on the United States for flour. The feeling in all the provinces is strongly in favor of closer free trade relations with the States. It has been asserted by some in the latter country, and by Canadians generally, that the present fiscal system of the Union has been ruinous to the nation, and that it tends to restrict trade and commercial intercourse between the United States and other countries to a very great extent. Viewed from a Canadian stand-point it is unquestionably ruinous to the Canadian Provinces. And there never was a time in the history of the latter when reciprocity with the United States was more urgently required than at present.

Since the union of the Provinces in 1867, there is a balance of trade, amounting to $229,696,886, against the Dominion; during the same time its public debt has been doubled; and the taxes, on account of the general government, rose from $13,486,032 to $24,488,372. The interest on the public debt is about seven millions of dollars per annum. And the state of the finances is such that additional taxes will have to be imposed in a short time. The taxes imposed by the general government are customs and excise. The amounts raised from these two sources ranged from $19,920,095 in the fiscal year 1874, down to $17,488,835 in the year 1877. The whole revenue for the year 1874 was $24,205,092; and in 1877 it was $22,059,274; being about one twelfth that of the United States. The exports of the Dominion for the latter year was about one-ninth that of the Union.

During the operation of the Reciprocity Treaty, the United States collected only about three-quarters of a million dollars on dutiable goods from the Provinces; while since the abrogation of the treaty the Dominion has paid from five to seven millions of dollars in gold annually, on the excess of imports from the States, into the United States treasury. Generally speaking, the duties are paid by the consumers. It is contended, however, by many in the United States that the quantities of her domestic products are so enormous, that the comparatively small amount of imports of like products from Canada have no effect in raising the price of such products in the markets of the Union.

It has been proposed in the Canadian Parliament to adopt a "National Policy," or in other words, a reciprocity of tariffs, as a means of compelling the United States to adopt a reciprocity of trade with Canada. There may be good policy in retaliations when two powers are about equally balanced in regard to extent and variety of available resources; but as regards the Dominion, such a policy could hardly fail to be ruinous to her own interests. Her sectional interests are so conflicting that she cannot adopt a system of protection which would be equally just to all the Provinces. Besides, a system of protection in Canada, similar to that in the United States, would retard reciprocity between these two countries for an indefinite period. Encouraged by the government, as in the States, capitalists would invest large amounts in certain industries, which could not fail to build up a powerful interest in opposition to reciprocity, as it has done in the Union. Men who have incurred vast expense on the faith of protective legislation, will justly claim that the government has no right to sacrifice their interests by a sudden change of policy.

There appears but little desire to renew the treaty of reciprocity on the part of the United States. Geographically considered, the States remote from the Dominion are opposed to reciprocity on account of their remoteness. Some of the border States are also opposed to closer trade relations with Canada. The agricultural, lumbering and coal interests are opposed to a renewal of the treaty. Indeed, all outside of strictly commercial localities, are united against it. The eastern States fronting on the Atlantic are most in favor. To the latter, at least, reciprocity could not fail to be highly beneficial.

In his address to Congress, in 1870, the President said: " No citizen of the United States would be benefitted by reciprocity. . . . The advantages of such a treaty are wholly in favor of the British Provinces." However, in 1874, at the request of Canada, the draft of a new treaty of reciprocity was prepared; but Congress refused to sanction it. Since that time but little effort has been made in favor of reciprocity. It is a question, however, whether reciprocity and union shall not mean the same thing in the near future.

Whatever may be said for or against the present commercial policy of the Union, the people of that country may congratulate themselves that their debt is being rapidly paid off, and hopes are entertained that it may be altogether extinguished before the close of the present century, notwithstanding the sad effects of their civil war and of the great depression in trade which followed. What would paralyze an old and densely populated nation of Europe has had but little effect on the United States, from the fact of its having enormous resources that cannot be exhausted. The present depression in trade in the United States is largely due to overproduction in her workshops. Indeed, Britain and other countries have suffered greatly from the same cause—overproduction of manufactured goods.

Having no colonies or eastern empire to defend, and separated from Europe by the Atlantic Ocean, the United States does not require either a large army or a navy. A nation numbering forty-five millions of inhabitants having a standing army of only 25,000 men, and whose navy requires only about 10,000 men and officers, is certainly not burdened in paying for their national defences.  Under its present systems, the United States can meet all its expenditure and have a large surplus for paying off the national debt.  While Great Britain spends about one hundred and fifty millions on her army and navy, it costs the United States less than sixty millions of dollars a year.

In old nation like Great Britain, which depends on other countries for the main part of its corn, population is apt to outgrow the means of subsistence, but in the United States, though the population doubles itself in half the time it does in Britain, the valley of the Mississippi alone could maintain the whole population of Europe.  And what is called the "balance of trade" is largely in favor of the United States.

The following table shows the value of the exports and imports of the United States and the Dominion of Canada for the years named:—

| STATE. | 1875. | 1876. | 1877. | 1878. |
|---|---|---|---|---|
| **UNITED STATES.** | | | | |
| Total Exports, gold value..........$ | 605,574,853 | 596,890,973 | 658,637,457 | 694,884,200 |
| Total Imports, gold value .........$ | 553,906,153 | 476,677,871 | 492,097,540 | 437,051,533 |
| Balance in favor of U. States ........ | 51,668,700 | 120,213,102 | 166,539,917 | 257,833,667 |

| DOMINION OF CANADA | 1875. | 1876. | 1877. | 1878. |
|---|---|---|---|---|
| Total Exports,....$ | 73,164,748 | 75,774,941 | 70,907,303 | .......... |
| Total Imports,....$ | 117,322,425 | 87,076,194 | 94,487,130 | .......... |
| Bal. against Canada | $34,157,677 | $11,301,253 | $23,580,827 | .......... |

Thus, in four years the United States had a total excess of exports over imports to the value of five hundred and ninety-six millions of dollars.  The increase in 1878 over the previous year was very large.  This is so much gold to aid the nation in resuming specie payments at the first of January, 1879, the time fixed by law.  The aggregate exports and imports of the United States increased from $613,000,000 in 1868, to $1,132,000,000 for the year ending 30th June, 1878.

Previous to the civil war, the revenue derived from customs, in the United States, was very small.  In 1840, it was only

thirteen and a half millions of dollars, being a little over half
that of the Dominion in 1878; while the population was ten
times greater than that of the latter is at present. In 1850,
the Union collected $39,668,686; and in 1860, $53,187,511.
But in 1870, in consequence of the civil war, it rose to $194,-
538,374; and in 1872, to $216,370,286. From the latter date,
it gradually fell, until in 1877, the total amount from customs
was only one hundred and thirty-six millions of dollars. The
net revenue of the United States for 1877, was $269,000,586;
and expenditure, $238,600,008; leaving a surplus of $30,340,-
577. The surplus of the previous year was $31,022,242.

The gold and silver products of the United States are enormous. In the "Report of the Commissioners of the General
Land Office for the year 1876," the total gold product of the
Union is estimated at $1,040,000,000, and silver product at
$100,000,000. It is not easy to ascertain the gold product of a
country. There is always more or less of which no account
is given. The amount for 1866 was set down at $57,000,000;
in 1867, at 70,000,000, and in 1878 to $47,226,107. Allowing,
however, fifty millions a year as the gold product of the
United States for the last ten years, then five hundred millions
of dollars worth of gold has been added to the wealth of the
country since 1868. In 1877, the gold and silver products
amounted to $80,533,064, of which $48,787,778 was gold; and
in 1878, to $103,952,421, of which $47,226,107 was gold. Of
the gold, California produced last year $15,260,679; Nevada,
$19,546,513; Montana, $2,260,511; Colorado, $3,366,404; Da-
kota, $3,000,000; Idaho, $1,500,000; Oregon, $1,000,000; and
small amounts were produced in eight other sections. The total
gold and silver product of the Union in 1878 and three previous years, was $331,000,000. Of the silver product, Nevada
produced $28,130,350; Utah, $5,208,000; Montana, $1,969,635;
Colorado, $5,304,940; Arizona, $3,000,000; California, $2,373,-
389; and small amounts were produced in six other localities.
The Director of the Mint estimated the value of gold, silver
and bullion in the United States at the first of October
1878, at three hundred and fifty-eight and a half millions of
dollars.

The total amount of banking capital of the United States
in November 1875, was $719,101,966, and total deposits
$2,036,296,106.

In the Dominion of Canada, the revenue derived from customs was $8,819,431 in 1868; $11,843,655 in 1871; $15,361,-382 in 1875; and $12,548,451 in 1877. The latter amount is about equal to that of the United States for the same year, according to population.

Referring to the last table, it will be noticed, that the Dominion imports are comparatively large according to population, being one-fifth that of the United States, while her exports are only about one-ninth that of the latter country. Hence, what is called the balance of trade is largely against the Dominion. In the three years named in the table, the excess of imports amounted to sixty-nine millions of dollars.

The trade problem is not at all times and in all countries an easy one to solve. There are ways in which a country may save itself from actual loss when the balance of trade is largely against it. Although Great Britain imports nearly one half the food of her people, her work shops supply the world with an immense amount of manufactured products, from which large profits are derived; and thus, in this way, as well as in many other ways, she is able to keep the finances of the nation fully recuperated, notwithstanding the balance of trade is against her.

The Dominion is differently situated. It has no variety of climate, except from cold to colder, and consequently no diversity of productions. Her chief settlements are scattered here and there, and far apart, along an east and west front, four thousand miles in extent. The interior of the country is shut out from the world by the United States in front. Hence, trade between the outer world and the interior and far west of the Dominion will always be a costly affair, unless with free and unrestricted trade with the United States. No two parts of the world have such excellent facilities for an interchange of commodities as these States and Provinces; and but few parts of the world have less commercial intercourse, owing to their relative positions, than these two countries. The Dominion has but few articles to export to Europe, but she has many that her own continent requires. Even her ships are shut out of the United States market; and her lumber, coal, salt, and other chief articles of export are subject to a high protective duty in the United States; hence the Dominion is obliged to send her chief exports to Britain at a small profit, if not, frequently, at a loss.

Goldwin Smith answers the question frequently asked— the question, why the Dominion had $23,500,000 against her in 1877, while the Union had $156,000,000 in her favor in the same year, and nearly two hundred and fifty-eight millions in the last fiscal year, when he says: "The only effectual remedy as I believe, is the removal of the customs line between us and the United States. What Canada needs is admission to the markets of her own continent, for want of which her lumber lies ruining its holders, and her manufacturers can scarcely keep themselves in existence." The gold product of the Dominion is comparatively small. Nova Scotia and British Columbia are the chief producers. In 1863, Nova Scotia produced $280,000 worth of gold. The total product between 1860 and 1875, was $4,610,953. British Columbia exported gold to the United States in 1863 to the value of $2,935,170. The export in 1876 amounted to $1,472,471; and to $1,188,739 in 1877. The total gold product of the Dominion up to 1878 is estimated at $66,768,000. The greater part of this amount has been sent to the United States.

The chapter of costs to Great Britain and the Provinces, of maintaining the International boundary intact, could not fail to be an interesting one. And since the union of the Provinces, this chapter has not been divested of interest. It costs Canada, directly and indirectly, an immense sum every year, and will no doubt continue to do so as long as it is dotted by a double line of custom-houses.

Labour generally follows capital. From one hundred and fifty thousand to two hundred thousands of the Canadian people remove to the United States every ten years. No matter what course Canada may pursue, her operations are limited by this boundary line. In consequence of the limitation of trade, and the increase of debt, the necessity for raising a larger revenue is becoming more and more pressing. While the taxes are being rapidly reduced in the States, in Canada new burdens will have to be imposed. With respect to population and resources, the public debt of the Dominion is comparatively enormous; and at the present rate of increase will soon become an intolerable burden to the country.

It is obvious, that unless trade between Canada and the United States shall be relieved of unnatural restrictions, and industry from unjust burdens arising from the construction of the Canadian Pacific Railroad, Canada cannot progress.

16

"Statistics," says Schlozer, "are history in repose," and "history is statistics in motion." The following statistical matter will be found to meet this view.

Table showing the exports of Canada to, and imports from Great Britain, United States, and other countries for the years named:—

| COUNTRY. | 1874. | 1875. | 1876. | 1877. |
|---|---|---|---|---|
| Exports to Great Britain .......... | $45,003,882 | $40,032,902 | $40,723,477 | $41,567,469 |
| do. United States. | 36,244,311 | 29,211,983 | 31,933,459 | 25,775,245 |
| do. other countries | 8,103,735 | 7,942,094 | 8,300,499 | 8,532,679 |

| COUNTRY. | 1874. | 1875. | 1876. | 1877. |
|---|---|---|---|---|
| Imports from Great Britain .......... | $63,076,437 | $60,347,067 | $40,734,260 | $39,572,239 |
| do. United States. | 54,283,072 | 50,805,820 | 46,070,633 | 51,312,669 |
| do. other countries | 10,044,660 | 8,465,770 | 7,928,925 | 5,415,575 |

Statement showing the excess of imports over exports by the Dominion, from Great Britain and the United States, for the years named therein:—

| COUNTRY. | 1874. | 1875. | 1876. | 1877. |
|---|---|---|---|---|
| From G. Britain ... | $18,072,555 | $20,314,165 | $9,983 | *$1,995,230 |
| " U. States ..... | 18,038,761 | 21,593,827 | 14,036,374 | 25,537,424 |

* In the year 1877, the value of the exports from Canada to Great Britain exceeded the imports from the latter country by nearly two millions of dollars. This arose chiefly from the increased export of the products of the forest, as the following table shews:—

| | 1876. | 1877. |
|---|---|---|
| Products of the Mine......... | $ 262,889 | $ 1,061,247 |
| "    " Fisheries.......... | 687,312 | 808,330 |
| "    " Forest.. .......... | 14,031,591 | 17,086,509 |
| Animals and their Products.... | 8,796,096 | 10,021,379 |
| Agricultural Products......... | 13,548,641 | 10,318,237 |

Statement shewing the value of the chief exports from Canada to the United States in each of the years named:—

| | 1874. | 1875. | 1876. | 1877. |
|---|---|---|---|---|
| Products of the Mine. | $3,611,607 | $3,487,968 | $3,201,588 | $2,447,844 |
| "   " Fisheries, | 1,705,813 | 1,644,828 | 1,475,330 | 1,317,917 |
| "   " Forest, | 9,871,749 | 6,694,746 | 4,973,354 | 4,789,594 |
| Animals and their products.. | 5,789,458 | 5,099,192 | 4,838,412 | 4,618,177 |
| Agricultural products, | 8,698,572 | 8,022,548 | 11,744,715 | 8,057,995 |

Table showing the value of dutiable and free goods imported into Canada and entered for consumption from Great Britain and the United States, for the years 1875, 1876 and 1877:—

| FROM GREAT BRITAIN. | 1875. | 1876. | 1877. |
|---|---|---|---|
| Dutiable.......... | $49,239,119 | $32,385,482 | $32,916,776 |
| Free.............. | 11,107,948 | 8,348,778 | 6,655,463 |
| Total.......... | $60,347,067 | $40,734,260 | $39,572,239 |
| FROM THE UNITED STATES. | | | |
| Dutiable.......... | $22,023,665 | $21,334,613 | $23,510,846 |
| Free.............. | 28,782,155 | 24,735,420 | 27,801,823 |
| Total.......... | 50,805,820 | 46,070,033 | 51,312,669 |

In 1873 Canada imported free goods from Great Britain to the value of twenty-one millions of dollars, while in 1877 less than one third of this amount sufficed.

The total value of the free goods imported from the United States into Canada for the five years ending with 1877 was $145,555,987, being about twenty-nine millions a year.

These figures clearly show that the United States is Canada's natural market to buy and sell in. In the evidence obtained by the Baie Verte Canal Commissioners in 1875, about fifty of the gentlemen who gave evidence were very decided in favor of reciprocity with the United States. What one said many others in substance said, that "the United States must inevitably be the great market for our surplus coal, plaster, freestone, grindstone, potatoes, oats, and various kinds of our lumber."

Statement showing the value of the imports from and exports to Great Britain and the United States, from and to the

Provinces therein named for the year 1877, designating the value of free goods imported:—

| | GREAT BRITAIN. | UNITED STATES. |
|---|---|---|
| Ontario Exports................ | $2,980,433 | $14,579,244 |
| " Imports................ | 11,724,328 | 28,193,329 |
| " Free Goods............ | 801,484 | 16,483,481 |
| Quebec Exports................ | $31,556,708 | $3,340,259 |
| " Imports ................ | 18,529,026 | 13,530,427 |
| " Free Goods............ | 4,323,525 | 7,486,145 |
| Nova Scotia Exports............ | $1,226,321 | $1,710,472 |
| " " Imports............ | 4,030,274 | 3,689,597 |
| " " Free Goods........ | 770,203 | 1,952,630 |
| New Brunswick Exports........ | $3,982,054 | $1,344,758 |
| " " Imports......... | 3,305,371 | 3,301,989 |
| " " Free Goods..... | 572,356 | 1,267,878 |

The exports from Quebec and New Brunswick to Great Britain were chiefly products of the forest. Of free goods, only $6,477,569 worth were from Britain; while free goods to the value of $27,190,134 were from the States.

### MERCHANT SHIPPING.

Statement showing the number and tonnage of vessels built; also, the total number and tonnage of sailing vessels, canal boats, and barges, owned by the United States and Canada in the years therein given:—

| UNITED STATES. | 1874. | 1875. | 1876. | 1877. |
|---|---|---|---|---|
| No. of Vessels built........ | 2,147 | 1,301 | 1,112 | 1,029 |
| Tonnage.................. | 432,725 | 297,638 | 203,585 | 176,591 |
| Total No. of Vessels owned. | .... .... | .... .... | 25,934 | 25,386 |
| Total Tonnage owned...... | .... .... | .... .... | 4,279,458 | 4,242,599 |
| DOMINION OF CANADA. | 1874. | 1875. | 1876. | 1877. |
| No. of Vessels built........ | 496 | 480 | 420 | 432 |
| Tonnage ................. | 190,756 | 151,012 | 130,901 | 127,297 |
| Total No. of Vessels owned. | 6,930 | 6,952 | 7,192 | 7,362 |
| Total Tonnage owned...... | 1,158,363 | 1,205,565 | 1,260,893 | 1,310,468 |

Thus, these two countries, in the last year named in the table, owned 32,730 vessels, measuring 5,634,476 tons.

Of the United States tonnage, sailing vessels and steam vessels measured 3,751,585 tons. Of the States, New York is the largest owner—having 1,280,637 tons; Maine, 522,948; Massachusetts, 485,408; and Pennsylvania, 380,320 tons.

Of the Provinces, Nova Scotia owned 541,579 tons; New Brunswick, 329,457; Quebec, 248,399; Ontario, 131,761; Prince Edward Island, 55,547; British Columbia and Manitoba, 3,725 tons.

When compared with the leading maritime nations of Europe, these comparatively young countries occupy an important position in regard to merchant shipping, as the following statistics show:—

In 1877, the tonnage of Great Britain, including her colonies, was 7,677,024; Norway, 1,391,877; Italy, 1,360,425; Germany, 1,053,229; and France, 870,225 tons. These figures do not include the inland tonnage of the sailing vessels of these nations, or of steamers under one hundred tons register, or barges.

The United States take rank as the second; and we feel safe in placing Canada as fifth amongst the ship-owning countries of the world.

Though Canada has no navy, still she is far ahead, in the tonnage of her merchant shipping, of the United States and of some of the great powers of Europe, in proportion to population. And her means of shipbuilding are still very great. However, the increasing demand for iron ships, and the large number being built in Great Britain and other parts of Europe, may before long lessen the demand for wooden ships.

## RAILROADS.

Railway construction was commenced in the United States in 1828. At the close of 1868, the system extended over an aggregate length of 42,255 miles. In 1876, the total length of the system was 74,095 miles.

In the Provinces, the first railway opened was in Canada proper, in 1847. In 1861, British North America had 2,162 miles constructed. And in 1876, the aggregate number of miles of railroad owned by the Dominion of Canada, was 5,494, exclusive of double track.

Comparing railway construction in these two countries with that in the great nations of Europe, the United States and Canada may well feel proud.

From the *Economiste Fraingaise* we learn that the German Empire at the end of 1876, had only 17,181 miles of railroad;

the British Islands, 16,794; France, 13,492; Russia, 11,555; Austria, 10,852; and Italy, 4,815; being a total of 74,689 miles of railroad, or only about six hundred miles more than the United States.

And the Dominion of Canada has nearly as great a railway mileage as the kingdom of Italy, and nearly half that of either Austria or Russia.

## COAL.

The carboniferous area in Nova Scotia and New Brunswick is comparatively large. With the exception of the *Albertite*, a highly valuable species of coal in Albert County, the coal of New Brunswick is not known to possess a high commercial value.

In Nova Scotia there is a surface of about 650 square miles, under which there are large deposits of excellent coal. The chief coal mines of this province are situated conveniently to places of shipment and railway transit.

Here is a source of wealth convenient to a great market. But the best market is all but closed. In 1855, when the effects of the reciprocity treaty were beginning to be felt, Nova Scotia exported coal to the United States to the value of $155,075, and in 1864, to the value of $684,642. From this date there was a decrease of shipments to the United States, arising out of the abrogation of the reciprocity treaty. The great demand for coal during the rebellion, and the depressing effects of that war upon productive industry in the United States, however, gave a great impulse to the coal trade; and in view of the abrogation of the treaty, efforts were made to force as much coal as possible into the markets of the Union before a duty should be imposed upon it. The exports fell from 540,744 tons in 1865, to 90,072 tons over imports in 1876; and to 61,037 tons over imports in 1877. This is a petty trade in coal between Nova Scotia and the United States. Had the reciprocity treaty continued in operation, the annual exports of coal from Nova Scotia to the States, would now, in all probability, be worth two millions of dollars.

## SALT.

This product, like that of coal, exists in numerous places in North America. In some places rock salt exists in solid masses near the surface; but most generally salt is produced by evaporation from brine springs, or from wells. The quantities annually produced are very large.

The United States in 1850 produced.......9,763,840 bushels,
Valued at.................. $2,222,745
In 1860, it produced.........................12,717,200 "
Valued at...................$2,289,504
In 1870, it produced.......................17,606,105 "
Valued at.................. $4,818,229

In the *Dominion of Canada*, the salt basin of Ontario is the only place where salt is produced to advantage. The quantity manufactured in 1873, was 451,576 barrels. Allowing a bushel to weigh 56 pounds and a barrel to contain five bushels, the product of that year was 2,257,880 bushels, nearly half of which was exported to the United States. In 1876, Ontario exported 870,437 bushels of salt to the United States. In 1877 the quantity exported was 785,973 bushels, valued at $81,443. Of the remainder very little was shipped to Canadian ports.

In his Geological Report for 1874-5, R. C. Selwyn, Esq., says: "The difficulties (arising from restricted markets) which beset salt-making in Canada, have not only prevented the boring of many more wells, but have checked production and improvements at wells already equipped and in action." Ontario exported 339,000 bushels less salt to the United States in 1877, than in 1873.

In 1877, Nova Scotia, New Brunswick, and Prince Edward Island imported 1,806,464 bushels of salt, valued at $182,548, from countries outside of the Dominion. However, the salt basin of Ontario and the coal beds of Nova Scotia are too far apart: it will not pay to exchange salt for coal; but the salt of Ontario is not far from the coal of Pennsylvania. Remove the custom-houses, and both salt and coal would find a ready market in the States. But even as it is, the inter-provincial barriers to trade are much greater than those of the international.

## CANAL STATISTICS.

To the Inland Revenue Report of Canada for 1877, we are indebted for the following figures.

Table showing the quantities of vegetable food carried to Tidewater by the canals and railways of the State of New York in the years named:—

| YEARS. | CANALS. | RAILWAYS. | TOTAL TONS. | PROPORTION BY CANALS. |
|---|---|---|---|---|
| 1869.... | 1,302,613 | 1,087,809 | 3,390,422 | .345 |
| 1871.... | 1,850,198 | 2,205,589 | 4,055,787 | .456 |
| 1874.... | 1,767,598 | 2,791,517 | 4,559,115 | .387 |
| 1876.... | 1,064,293 | 2,875,803 | 3,940,096 | .270 |
| 1877.... | 1,498,984 | 2,493,683 | 3,992,667 | .375 |

Statement showing the tonnage of vegetable food carried on each of the lines of Canals, and on the two principal railways competing for the carrying trade between Lake Erie and Tidewater for the years named:—

|  | 1869. Tons. | 1871. Tons. | 1874. Tons. | 1876. Tons. | 1877. Tons. |
|---|---|---|---|---|---|
| Total on New York Canals. | 1,302,613 | 1,850,198 | 1,767,598 | 1,064,293 | 1,498,984 |
| Total on Welland Canal... | 503,860 | 668,076 | 622,558 | 455,022 | 406,567 |
| Total on N. Y. Central and Erie Railways, | 1,087,809 | 2,205,589 | 2,791,517 | 2,875,803 | 2,493,633 |
| Quantity cleared at Buffalo and Tonawanda by Erie Canal.... | 786,436 | 1,315,693 | 1,157,509 | 783,331 | 1,223,100 |
| Quantity cleared at Oswego by Canal........ | 267,815 | 297,424 | 243,325 | 99,975 | 126,899 |
| Quantity cleared through the Welland Canal in transit between ports in the U. S...... | 337,530 | 384,585 | 290,114 | 181,885 | 169,836 |

A glance at these tables clearly shows that railway transit, in consequence of its superior speed, is preferred, for flour and other vegetable food, to that by canals. Hence, a question arises whether it will pay to make the contemplated enlargement of the Erie Canal; and whether the enlargement of the Canadian canals, which have already cost many millions of dollars, will be worth half the expenditure. In 1869, according to the previous table, more than half was conveyed by canal; while in 1877, a million tons more vegetable food was conveyed by the railways than by the canals of New York. And by the last table, it will be observed the Central and Erie Railways succeeded in increasing very largely the proportion carried by them; notwithstanding that, in 1872, the tariff on wheat, barley, rice, anthracite coal and iron ore was reduced one-half; on corn and oats two-fifths; and on railway iron, domestic coal, and bituminous coal, one-third, *via* the New York canals.

Though canal transit was generally considered the most favorable means of conveying heavy goods, such as iron, salt, and coal, still the quantities conveyed by this means does not seem to increase. The report says: "The quantities of vegetable food passed through the Welland Canal (Canadian) in transit between ports in the United States has largely decreased, . . . and the decrease in 1877 is greater as compared with 1869 than in any preceding year. There has also been a decrease in the quantities of heavy goods."

That the great network of railways, constructed and yet to be constructed on this continent, will be the principal lines of transit for nearly all kinds of goods, there is now but little doubt. The age is a fast one; transit by lakes, rivers and canals is too slow to meet its requirements.

### POSTAL AFFAIRS.

There is no department of the public service of so great importance to society as that of the postal affairs of the country. This will more fully appear by reference to the following statistics for 1877:—

#### UNITED STATES.

| | |
|---|---|
| The number of ordinary Stamps issued was | 689,580,670 |
| Newspaper and Periodical Stamps | 1,388,709 |
| Stamped Envelopes, plain | 84,285,700 |
| Stamped Envelopes, request | 64,374,500 |
| Newspaper Wrappers | 21,991,250 |
| Postal Cards | 171,015,500 |
| Official Postage Stamps | 18,867,145 |
| Official stamped envelopes and wrappers | 14,750,445 |
| Aggregating | 1,060,253,919 |

17

| | |
|---|---|
| Number of Post Offices | 37,345 |
| Length of Mail Route, miles | 292,820 |
| Number of miles travelled | 147,353,251 |
| Including No. of miles by railroad | 74,546 |
| Steamboat Routes, miles | 17,685 |
| Number of Money Order Offices | 4,144 |
| No. of Domestic Money Orders | 4,925,931 |
| Payments by Domestic Money Orders | $72,908,475 |

Revenue .................. $34,544,885
Expenditure .............. 33,486,322

In favor of the Department ... $1,058,563

DOMINION OF CANADA.

| | | |
|---|---|---|
| Number of Letters | | 41,510.000 |
| " | Post Cards | 5,450,000 |
| " | Registered Letters | 1,842,000 |
| " | Newspapers and Periodicals | 39,000,000 |
| " | Books and Miscellaneous Articles | 4,638,000 |
| " | Parcels | 90,000 |

Aggregating .................. 93,626,000

| | |
|---|---|
| Number of Post Offices | 5,161 |
| Miles of Mail Route | 38,526 |
| Including distance by Railroad | 4,576 |
| Number of miles travelled | 15,126,676 |
| Number of Money Orders | 758 |
| Amount of Money Orders | $6,856,837 |

Revenue .................. $1,501,134
Expenditure .............. 2,075,618

Against the Department ...... $574,484

Articles of agreement have been concluded between the United States and British North America, providing that all money orders mailed at the exchange offices in each country and addressed to payees shall be transmissible in the mails free of postage.

The Official Report of the United States for the fiscal year ending June 30, 1877, says: "The number of orders issued in the United States, payable in the Dominion, was, 10,768, amounting to $227,216, and the number of Canadian orders paid in the United States was 16,231, amounting to $297,838. The amount of Canadian Orders issued in the United States and afterwards repaid was $1,167."

The Canadian Official Report for the same year, says: "The exchange of money orders with the United States during the year has been as follows: Amount issued, $277,969; and the amount paid, $208,133." The United States Report says, that "a comparison of this business with that of the previous year shows an increase of $40,220.48, or 21.51 per cent. in the amount of orders issued; of $65,221.43, or 28.03 per cent. in the amount of orders paid."

## REPRESENTATION.

It must be obvious to every student of Canadian history that the British North America Act was framed under circumstances highly adverse in many respects to the true interests of the Dominion of Canada.

During the discussion of the Confederation platform, a strong opposition to the Union prevailed in all the Provinces, and especially in Canada East and the Lower Provinces. Hence, in order to reward the leading advocates for union, as well as the prominent opponents of confederation, offices of endowment were multiplied; titles of honour were conferred by the Queen on half a dozen Canadians; and indeed, places were provided both in and out of parliament for a legion of office-holders. Had the constitution been framed irrespective of monarchical predilections, and with a due regard to the interests of American society and the resources of the country, such a costly system of government would not have been imposed upon this young community.

We look in vain to find its counterpart. Let the honours conferred go for what they are worth, and that is not much in American society. But here we have a Central Parliament, clothed with all the senseless trappings of royalty, and seven local legislatures, making laws for a population about equal to that of the single State of Pennsylvania; and this we have to pay for at a great price.

Economy was no part of the purposes of the delegates who framed this Act. The United States framed their constitution in their own country, irrespective of monarchical extravagance. Though the Canadian delegates knew that the salary of the President of the United States was only twenty-five thousand dollars per annum for presiding over thirty-six millions of people, they agreed to give forty-eight thousand seven hundred dollars a year, as a salary to one of Great Britain's lords to preside over three and a half millions of Canadians,

It was not until 1874, when the United States contained about forty-three millions of inhabitants, that that nation gave its chief magistrate a salary of fifty thousand dollars a year. Thus, during seven years, the Dominion paid its Governor-General $100,000 more than the President of the United States received in that time. Besides the travelling charges and contingencies of office of the Governor-General are nearly double that of the President of the United States. And the good-natured tax payers of the Dominion pay it, but not without a murmur.

Some years ago the Central Parliament took steps to reduce the Governor-General's salary to thirty-two thousand dollars a year; but the Imperial Government vetoed the measure. Evidently in this and other ways, the Dominion is paying the full price for British connection.

Canada may boast of having the most prodigal system of government in the world. The old thirteen colonies, with a population at the time of the revolt about equal to that of the Dominion in 1878, began with only twenty-six senators. The Dominion has 76, and 206 members in the House of Commons.

At present the United States Congress is composed of 77 senators, 291 representatives, and a delegate from each of the eight territories—in all 376; while the Dominion can boast of 282, besides 391 in the local legislatures; that is, 673 men making laws for about as many people as there is in Ireland.

Though each state of the Union is largely represented in the local legislature, the number is not so great nor the cost so large, according to population, as in the Provinces.

At the rate of increase established by law, if ever the Dominion contains one-fifth as many people as now inhabit the United States—say nine millions—the House of Commons of Canada will contain upwards of 450 members; and on the same principle, the House of Representatives in the United States would now number about 2,800 members. But the Union started on a different principle, and certainly has no need to regret it.

And what is still more to be regretted with regard to the Dominion, the people have no power in the matter without appealing to a power residing three thousand miles from their shores.

Ireland has only 105 members in the British Parliament, and Scotland 60. And these numbers have remained the same, or nearly so, for more than a life time. The State of Pennsylvania has two senators and twenty-seven representatives in Congress, and New York, containing 650,000 more inhabitants than the whole Dominion, has only two senators and thirty-three representatives in Congress.

The following statement will be found useful for comparison:—

|  | MEMBERS. | POPULATION. |
|---|---|---|
| British House of Commons.... | 658 | 33,500,000 |
| Italy............................ | 503 | 27,000,000 |
| Prussia ....................... | 431 | 26,000,000 |
| Austria....................... | 353 | 21,000,000 |
| Hungary..................... | 351 | 16,000,000 |
| Spain......................... | 336 | 17,000,000 |
| Belgium ..................... | 120 | 5,500,000 |
| United States................. | 291 | 45,000,000 |
| Canada. ..................... | 206 | 4,000,000 |

Except the United States and Canada, the above countries have each only one legislative body, composed of two branches. But in no part of the machinery of Government is the Dominion more extravagant than in regard to the Executive. In framing the Constitution, knowing that each local Legislature had an Executive Council, it might, in all reason, have sufficed to have limited the Governor General's ministry to eight members, the number of the President's Cabinet in the United States. But in Canada no less than thirteen ministers are drawing $91,000 annually from its revenues, besides supporting a vast amount of costly machinery attached to each department. At first each minister received $5,000; it was afterwards increased to $7,000 per annum. Shortly after the union of the Provinces the Hon. Alexander Mackenzie said: "The United States, a country with 40,000,000 of people, is governed by seven ministers, while we, with 4,000,000 of people, require thirteen. It is absurd and monstrous. There the Secretary of the Treasury attends to all the financial business. Here the Finance Department is divided up between the Finance Minister, the Receiver General, the Minister of Customs, and the Minister of Inland Revenue, four of them." He adds: "The system is extravagant in the extreme;" and he might with justice have said that the whole system of central legis-

.la'ion and government in Canada is a gross imposition on a generous-hearted people. The chapter of costs connected with this Parliament is a remarkable one for extravagance, especially when compared to the infancy of the country, to the numbers governed, and to the central legislature of the United States. But as we stand there is no hope for retrenchment.

## FORESTS.

North America is losing her forests rapidly, and consequently is likely to suffer greatly in many of her industries. Public attention has been officially called to the subject on both sides of the international line. In his report for 1876, the Commissioner of the General Land Office of the United States says, p. 7, that "a national calamity is being rapidly and surely brought upon the country by the useless destruction of the forests. Much of this destruction arises from the abuses of the beneficent laws for giving land to the landless." But the difficult task is to recommend what ought to be done to preserve the forests from waste, and for future use. In his report for the following year he says: "All past history shows only two successful methods of preserving timber in densely populated countries; the one by the Government retaining the title to the land and exercising a watchful supervision over the sale and disposition of the timber, as in Germany, where large revenues are annually derived from this source; the other through law of entail, as in England, by means of which a landed aristocracy holds the soil, and has the aid of the strong and well executed laws in securing the preservation of the timber." And the President, in his last message, has earnestly drawn the attention of Congress to the subject. And in his last annual report the Secretary of the Interior says: "The rapidity with which this country is stripped of its forests must alarm every thinking man. It has been estimated by good authority that if we go on at the present rate the supply of timber in the United States will, in less than twenty years, fall considerably short of the home necessities. It is the highest time that we should turn our earnest attention to this subject, which so seriously concerns our national prosperity."

And the Commissioner of Agriculture, in his Report for 1875, has devoted 115 pages of that work to the consideration of the subject. He says: "Forestry has excited much attention in the United States in recent years, in consequence of the rapid deforesting of large areas, and the expression of fears of a timber famine at no distant day. That the great white pine forests are being rapidly despoiled of their original growth, and that inroads are being made upon the heavy timber of the sierra slopes and deep valleys, there can be no question; and yet there is much that is sensational and extravagant in the views of alarmists on this subject." He says: "The western slopes of the sierras are prolific of new growths in place of the old, and, except in the vicinity of the Central Pacific road, are almost untouched by the woodman's axe, as also are the immense forests of Washington Territory and Oregon. More than half the entire area of the south is woodland." He estimated the total forest area of the United States and Territories at twenty-five per cent. of the total area of the Union. This is far below that of Norway, Sweden and Russia, which the German writer, Reutzeh, says is respectively 66.60 and 30.90. Germany comes next at 26.58; France, 16.79; and so on downwards to Britain, 5, and Portugal, 4.40 per cent.

And in an official report for 1878 it is clearly shown, that in the short space of the last two or three years the destruction of United States forests has been immense; indeed, much greater than in the comparative past. It is only necessary to quote the exports of wood of all kinds, which does not exceed the value of fourteen millions of dollars per annum, to show how scarce useful wood is becoming in the Union.

And in the Dominion of Canada a similar alarm has been sounded. In his report for 1875, the Surveyor General of New Brunswick says that "the public are as yet hardly aware how fast our good timber lands, available for revenue purposes, are decreasing in extent—decrease caused more by fire than by the axe of the lumber operator; both, however, fast destroying the great wealth of the Province." Referring to some of the principal lumbering regions of the Province, he says: "Many rivers emptying into the Bay of Fundy, which formerly yielded great quantities of spruce lumber, now produce none. The Grand Lake country, from which for years more than sixty millions of spruce and pine logs were brought to market, did not last year, nor will it during the coming

year, yield much over ten millions." He names half a dozen
other large districts which "have dwindled down to probably
less than one tenth of their former product."

H. G. Joly, Esq., in the last Report of the Minister of Agriculture, says: "Thinking men have begun to sound the note
of alarm." He calls our "attention to the pine and spruce, as
they form nearly all our export to Europe, and are really the
produce of our forests; while the hardwood we export, especially the fine oak, nearly all comes, at present, from the
Lake regions of the United States, as we have very little of our
own left."

"For some time past the idea has been gaining ground
among men who take an interest in the future of the country,
that our great pine and spruce forests are getting rapidly exhausted, and that, before long, a trade which enables us to
export annually over twenty millions of dollars worth of
timber (nearly twenty-seven millions worth in 1874, twenty-
five millions in 1875, and twenty millions three hundred
thousand in 1876), will shrink down to wofully reduced proportions." After naming the chief lumbering regions in
Canada proper and in New Brunswick, where operations have
been carried on for a long time, and after showing that these
regions are denuded of a large part of their best forest wood,
especially pine, the report says with regard to "that huge
tract of lumber country between the Ottawa and the St.
Maurice, that separated (or rather appeared to separate) the
lumbermen working on these two rivers by what seemed an
inexhaustible and endless forest, that huge tract is tapped
through and through, and the Ottawa lumberman has met the
St. Maurice lumberman on the shores of Lake Manooran. A
glance at the map will show what that means. Those who
think that there will never be an end to our timber may say:
'We can still go north.'" To this he replies: "Not very far
north." The distance is too great, "and the country is generally poor and barren."

"In a very short time, since the beginning of this century,
we have overrun our forests, picked out the finest pine, and
we have impoverished them to a serious extent, and what
makes it worse, impoverished the country too, for, owing to
the force of circumstances which we shall consider later, our
timber export trade has not given Canada such a return as she
had a right to expect. There still remains to us a great deal

of spruce and second-rate pine, which for generations to come will be in excess of our local wants if we are careful; but the *really fine pine* required to keep up our great timber export trade to its present standard is getting very scarce and inaccessible, and I fear that we must prepare for a sudden and considerable falling off." And the soil of the great pine regions of North America is not generally favorable to agriculture.

Of the Atlantic Provinces New Brunswick is the only section which can furnish any considerable quantity of lumber; and the quantity is being rapidly diminished. Nova Scotia has but little to spare, and Prince Edward Island has not enough for domestic wants. There is a tract of good timber land in the interior of Newfoundland, but unless it shall be connected with the sea-board by railroad it cannot be made available.

And in the Province of Quebec "the old settlements are painfully bare of trees." The report says : "You can sometimes go miles without seeing any tree worth looking at, and the passing stranger fancies himself in a country more denuded of trees than the oldest parts of Europe. There is a large district of very good agricultural land south of Montreal, where the scarcity of firewood, which is a matter of life and death in our climate, has compelled many a farmer to sacrifice a fine farm and leave the country. There are many other spots in the Province nearly as bad, and unfortunately the process of destruction is going on even now in more places than one."

And large sections of the Province of Ontario are denuded of forest wood; consequently this Province is a large importer of coal from the United States.

"We can cope," says Mr. Joly, "with waste and pillage in our forests; they are but the work of men, but we are terribly helpless against fire. . . . It is estimated by those who are most competent to form an opinion on the subject, that more fine timber has been destroyed by fire than has been cut down and taken out by the lumberman."

Against the destruction of our forests by fires set by lumbermen, hunters, Indians, and others, the laws are powerless. And the fires of settlers have frequently made great ravages. The area of country laid waste by fire in the Provinces and States is truly immense.

"We have not," says the report, "been spending the income or annual profit of our forests, but the forests themselves— not the interest but the capital."

18

The difficulty, however, is to find a practicable remedy. In his last report, the United States Commissioner recommended, "that Congress be requested to enact such laws as may be necessary for the appraisement and sale of such timber lands as it may deem best to sell; also to provide for specific legislation for fines and punishment for trespass on the timber on all public lands."

Mr. Joly "recommended limiting the lumberman to a maximum cut of so many thousand feet per square mile of his limits."

However, it is not only the scarcity of wood and the destruction of a great source of revenue that are to be deprecated, but the "disastrous climatic effect resulting from the removal of forests." It is generally allowed that one-third the area of a country should be in a forest state. Forest culture is largely encouraged in many of the old States of Europe. And on the western plains of North America forests are being planted.

Though our limited space forbid further amplification, we are unwilling to dismiss this important subject without special reference to the Dominion.

The reader, in comparing the productions of the Provinces with those of the States cannot fail to realize the necessity on the part of Canada, to protect and economise every source of revenue she possesses. The Dominion is not an agricultural country, but she has immense forests from which large revenues have been derived. Destroy the forests, and the great staple of the country will soon be among the things of the past. Already the exports of pine and hardwood are matters of history. Spruce is the only forest wood on which the Dominion can rely, at least to any great extent. And it has the advantage of reproducing itself at a rapid rate. But the destructive agencies at work point to a near future when this source of revenue will fail also, unless forcible measures shall be immediately adopted to prevent it. The preceding extracts from reliable authorities point to this fact.

During the last forty years the writer of these pages has had many opportunities of noting the agencies by which large sections of Nova Scotia and New Brunswick have been divested of their most valuable timber. As a land surveyor, during that time he has traversed the country in various directions; besides having read most all the available reports of public surveys for railroads and other purposes. From the

facts adduced, we are led to believe that the destruction of
the forests of this country is much greater than has been
generally estimated. And what is true of these two Provinces is, no doubt, equally true of large sections of the other
Provinces.

Looking at the extent of our forest land the casual observer
might conclude that we still possess inexhaustible forest resources. But such is not the case. Large regions of the
forest lands in all the Provinces and in the adjoining States,
are utterly worthless either for settlement or for timber. Still
we possess vast areas of forest lands which are continually
reproducing large quantities of useful timber.

However, in place of wasting so much wood in the manufacture of deals and square timber, as we are doing; in place
of selling the products of our forests in the markets of England for less than half the real value, we should limit the
annual production to less than half that of the present.
Norway and Sweden, which have continued to supply the
markets of Europe with immense quantities of forest products for the last hundred and fifty years, have long ago
taken steps to prevent the destruction of their forests; besides,
they economise by manufacturing all kinds and sizes of
wood, and making it ready for the various uses for which it
is intended in the markets. They do not leave a large part of
every tree, as is done in this country, to rot in the wilderness;
they do not waste one-third of their pine by manufacturing
a part of each tree and leaving the remainder of the trunk as
fuel for the flames.


DISTANCES.

The following table of distances, compiled chiefly from
official sources, will be found useful:—

|  | Miles. |
|---|---|
| Halifax to Liverpool | 2,482 |
| " Pernambuco, Brazil | 3,331 |
| " St. Thomas | 1,630 |
| " Esquimault, Pacific Ocean, by nearest land route in Dominion | 3,986 |
| " Moncton, by Intercolonial Railroad | 188 |
| " Restigouche " " | 386 |
| " Riviere du Loup " " | 562 |
| " Quebec, by Intercolonial Railroad | 682 |
| " Montreal, " " | 840 |

|  | Miles. |
|---|---|
| Halifax to Toronto, by Intercolonial Railroad | 1,180 |
| " St. John, N. B., " " | 276 |
| " Bangor, by railroad | 478 |
| " Danville Junction | 588 |
| " Montreal, via Danville | 858 |
| " Boston, by railroad | 723 |
| " do. by water | 364 |
| " New York, by railroad | 956 |
| " do. by water | 540 |
| " Windsor, by railroad | 171 |
| " Picton, " | 113 |
| " Cape Race | 463 |
| " San Francisco, via Quebec | 4,021 |
| " Victoria, Pacific, via City of Ottawa | 3,870 |

| St. John to Liverpool | 2,683 |
|---|---|
| " St. Thomas | 1,680 |
| " Pictou, by water | 480 |
| " Montreal, " | 1,180 |
| " Cape Race | 715 |
| " Charlottetown | 493 |
| " Toronto, by Intercolonial Railroad | 1,084 |
| " Boston, by railroad | 447 |
| " New York, " | 680 |
| " Montreal, via Bangor | 582 |
| " do. via Intercolonial | 740 |

| Montreal to Quebec City, by water | 160 |
|---|---|
| " West Light, Anticosti | 545 |
| " Straits of Belle Isle | 986 |
| " Cape Race | 996 |
| " St. Thomas | 2,145 |
| " Pernambuco, Brazil | 3,956 |
| " Havanna | 2,598 |
| " St. Thomas | 2,430 |
| " Rio Janeiro | 5,330 |
| " New York | 459 |
| " Boston | 279 |
| " Liverpool, by Cape Race | 2,969 |
| " do. via Belle Isle | 2,682 |
| " City of Ottawa, by water | 116 |
| " Kingston, " | 178 |
| " Port Dalhousie, " | 348 |
| " Duluth, Lake Superior, by water | 1,398 |
| " Chicago, " | 1,261 |
| " Portland, by railroad | 297 |
| " Boston " | 334 |
| " New York, " | 404 |
| " Chicago, " | 847 |
| " Omaha, " | 1,249 |
| " San Francisco, " | 3,181 |

| Quebec to Portland | 317 |
|---|---|

| Toronto to Quebec, by railroad | 555 |
|---|---|
| " Halifax, " | 1,237 |
| " Boston, " | 522 |
| " New York, by railroad | 546 |
| " Fort Garry, via Chicago | 1,537 |

| | Miles. |
|---|---|
| Toronto by railroad to Winnipeg, via Detroit and Chicago.... | 1,589 |
| Pictou to Montreal, by water............................. | 860 |
| " do. by railroad..................... .... | 83) |
| " Toronto, by railroad........................... | 1,227 |
| " Boston, by water............................... | 600 |
| Prince Edward Island Railroad........................... | 198 |
| City of Ottawa to Bute Inlet, Pacific, via Canadian Pacific Railroad line............................ | 2,774 |
| Head of Bute Inlet to Pacific Ocean...................... | 160 |
| Ottawa River to Red River, via north shore of Lakes Huron and Superior............................... | 1,200 |
| Lake Superior to Red River, Canadian P. R. Line............ | 410 |
| " " Burrard Inlet, Pacific..................... | 1,973 |
| Victoria, Vancouver Island, to Sidney..................... | 6,895 |
| " to Amoor......................................... | 3,900 |
| " to Shanghai...................................... | 5,220 |
| " to Canton....................................... | 5,980 |
| " to Melbourne.................................... | 6,940 |
| Duluth to Fort Garry, via United States.................... | 455 |
| Fort Garry to Edmunton, by water ....................... | 1,150 |
| Sorel River, St. Lawrence, via Lake Champlain, by water to Albany, New York.............................. | 411 |
| Albany, to Buffalo, via Erie Canal......................... | 350 |
| New York to Liverpool.................................... | 3,013 |
| " to Jamaica .................................. | 1,530 |
| " to St. Thomas................................ | 1,420 |
| " to Barbadoes...... ......................... | 1,800 |
| " to Pernambuco, Brazil ........................... | 3,364 |
| " to Cape Race, Newfoundland................... | 1,010 |
| " to Charlottetown ............................ | 780 |
| " to Chicago via Erie Canal and Lakes Erie and Michigan ............................... | 1,504 |
| " to Queenstown ............................ | 2,773 |
| " to Philadelphia, by railroad..................... | 90 |
| " to Chicago, " ................... | 913 |
| " to Omaha, " ................... | 1,405 |
| " to Cheyenne, " ................... | 1,932 |
| " to Ogden, " ................... | 2,435 |
| " to San Francisco, " ................... | 3,317 |
| Chicago to Liverpool, via Mississippi River................. | 6,000 |
| " to " via Erie Canal and New York........ | 4,000 |
| " to " via Welland Canal and St. Lawrence.. | 4,160 |
| " to Buffalo, by water............................. | 916 |
| " to Quebec, " ................... | 1,421 |
| New Orleans to Jamaica.................................. | 1,095 |
| " to Barbadoes ............................. | 2,120 |
| " to St. Thomas............................ | 1,630 |
| " to Havanna.............................. | 590 |
| Mississippi River and Branches is navigable for............ | 20,600 |
| Length of Mississippi..................................... | 3,160 |
| " to head of Missouri Branch........................ | 4,491 |
| Ohio River, navigable for................................. | 1,175 |
| or, 2,000 miles from Gulf of Mexico. | |
| Cape Race, Newfoundland, to Cape Clear, Ireland........... | 1,640 |

142

### LAKES.

Lake Ontario is 180 miles long; mean breadth, 65 miles; mean depth, 500 feet; 262 feet above the sea; and area, 7,000 square miles. It is 756 miles from the Gulf of St. Lawrence.

Lake Erie is 240 miles in length, and 80 in breadth; 84 feet mean depth; 555 feet above the sea; and area 10,000 square miles.

Lake St. Clair is 20 miles long, and 36 wide; 20 feet deep, and 571 feet above the sea; area 360 square miles.

Lake Huron is 260 miles long, by 160 wide; mean depth, 800 feet; 574 feet above the sea; and contains 20,000 square miles.

Lake Michigan is 360 miles in length; 108 miles in width; 800 feet deep; 587 feet above the sea, and contains 20,000 square miles.

Lake Superior is 355 miles in length; 160 in width; 988 feet deep; 600 feet above the sea, and covers 40,000 square miles.

Lake Champlain is below these lakes. It is 110 miles long; greatest width, 14 miles; and depth from 50 to 280 feet. It discharges by Richelieu River into the River St. Lawrence, about 45 miles below Montreal.

## POLITICAL UNIFICATION.

The history of the world, especially that of Europe, marks the dismemberment of many ancient States, the union of some, and the partition of others, out of which great powers have been erected.

Lord Houghton said, on seconding the address in reply to the speech from the British throne in 1867, that "the future of the world rested not in insulated municipalities, but in great empires." And the Marquis of Salisbury, in a recent speech in the House of Lords, said: "That the small kingdoms which we have guaranteed are inviolably doomed to *evital* destruction. Almost every generation sees the absorption of one or other of them. The future is one of great empires, and small powers will have hard work to live at all." Sir C. W. Dilke, in his *Greater Britain*, p. 199, says: "It is small powers, not great ones, that have become impossible."

In no part of the world is this doctrine more completely realized than in the British Islands. The seven kingdoms of England, known as the Saxon Heptarchy, were originally independent of each other. Several times however, a powerful Sovereign acquired a preponderating influence over the other six nations. Scotland and Ireland were each divided into a number of petty nations. France, at one time, was divided into several nations. Germany, says Russell, "was formerly possessed by a number of free and independent nations, . . . extending from Bohemia to the Baltic and the German Ocean." Out of this cluster of States Austria and Prussia sprang. The former became a great power; and, until a recent date, was head of the German confederation.

Prussia, in 1815, only contained ten millions of inhabitants. At the beginning of her last war with France her population was about twenty millions; before the close of that war, by acquisitions of neighbouring states, Prussia numbered more than forty millions.

In his History of the Middle Ages, Henry Hallam says: "The German Empire indeed had now assumed so peculiar a character, and the mass of States who composed it were in so many respects sovereign within their own territories, that wars unless in themselves unjust could not be made a subject of reproach against them." And "The republics of Italy in the thirteenth century were so numerous and independent, and their revolutions so frequent, that it is a difficult matter to avoid confusion in following their history." But it remained for the latter half of the nineteenth century to organize the kingdom of Prussia into an empire, and consolidate the Italian States into a kingdom.

We might search the whole compass of European history in vain to find such important results as followed the short wars of 1866 and 1870. They closed in laying the foundation of an united Germany, and in the unification of the Italian States. Austria was compelled to cede to Hungary the chief part of her historic rights, and let go her grasp upon Italy. And France, by means of her war with Prussia in 1870, was compelled to withdraw her forces from Rome, and leave Italy to the King of Sardinia. In this war "we see," said an eloquent writer, "one Germany rising in its strength, gathering in its avalanche of excitement all its manhood to battle, all its old age to guard, and all its womanhood to tend and heal the wounded and the sick." Thus Prussia was introduced into the family of great powers; Italy became a strong nation, with Rome as its capital; France, though humbled, is one of the great powers; and the Pope's civil power was terminated.

And at the close of the Russian and Turkish war in 1878, followed by the Berlin treaty, the map of the East was again changed. Great powers were made greater and feeble States were enfeebled. Russia, which is the unification of a group of Slavic and other States, continues to march on by large strides. Turkey, by the common consent of the Berlin Congress, was partitioned; and Russia gains the portion of Bessarabia which she lost by the treaty of 1856, and thus once more extends her. frontier to the Pruth and the Danube. The area of this accession is upwards of 3,300 square miles. And on the Asiatic frontier, at the eastern end of the Black Sea, Russia obtained an accession of territory containing about nine thousand square miles. These two areas contain about 78,000,000 acres. And Batoum also is annexed to Russia, with the understanding that no fortifications shall be maintained on it.

Bosnia and a part of Montenegro were ceded to Austria. The town and valley of Kotur, which lie to the east of Lake Van, are restored to Persia.

Servia, Roumania, and the remaining part of Montenegro are independent, and Bulgaria is practically so. All these with Greece are feeble States; and Turkey is so weakened by war, together with the loss of about 71,500 square miles of her domain, that she is powerless except for murderous oppression.

Thus the troubles in the East, and they are many and complicated, have been settled, for the present by the Treaty of Berlin. Never did peace speak with a more feeble voice. If this Congress had annexed all the feeble states to great powers, some hope might be entertained of a more lasting peace among the Eastern Nations. But as matters stand, if Russia, Austria, and other great powers aim at a reconstruction of their frontiers—if they aim at extending their boundaries, the way is open for the annexation of feeble states. This may not be effected without bloodshed, as in the past, unless the great powers convene another Congress for the purpose of annexing them.

Evidently the destiny of feeble nations is absorption by great powers. "Small powers will have hard work to live at all." Poland, Hungary and Bohemia were dismembered.

Alliances may save weak nations for a time. The Netherlands and the seven Provinces of Holland, Norway and Sweden have their unions.

Spain, before the middle of the eleventh century, was divided among a number of independent sovereigns, whose history became less important in proportion to the increase of the number of kingdoms. Ultimately the small states of the Spanish Peninsula united, and Spain became a great power, whose dominion extended over Gibraltar, the Netherlands, the Two Sicilies, and an immense empire in America. Her boast was that "when Spain moves the whole world trembles." Through a conjuncture of events, however, she lost nearly all her distant possessions. The few feeble nations that remain seem to be mere reservations until the great powers are ready to change their political destiny.

The effect of the consolidation of the world into great powers can hardly fail to be salutary. The number of disputants, and consequently the number of national disputes are

19

decreasing. Those nations long at war with the aspirations of freedom are greatly changed, and higher aspirations, motives and ideas—a higher standard of moral and intellectual progress more becoming the manhood of nations, is assumed. The dissolution of church and state and the gradual adaptation of national constitutions to the wants of society are taking place among the great powers. Greater freedom of conscience, of discussion, and of the press, is leading to a more peaceful development of the resources of each nation than existed among the small nations of the past. Consequently we are enabled with much less difficulty than formerly to assign the positions, and outline the future of nations. No limit has been set to the extent of empire, number of people, and variety of interests which may be governed by one executive head. The world has long had the experience of government by absolute and despotic monarchs whose will had the force of law. Such were the governments of China, Russia, Austria, Spain, Turkey, France, and other nations.

At the time of the discovery of America by Columbus, and long after, the moral tone and intellectual character of the most enlightened nations of Europe were at a low ebb. Even in the British Islands civilization was low, and liberty of conscience was looking abroad for a safe resting place. "The liberties of our country," says Macaulay, "were in the greatest peril. The opponents of the government began to despair of the destiny of their country, and many looked to the American wilderness as the only asylum in which they could enjoy civil and spiritual freedom. There a few resolute Puritans who, in the cause of their religion, feared neither the rage of the ocean nor the hardships of uncivilized life, neither of savage beasts nor the tomahawks of more savage man, had built amidst the primeval forests villages which are now great and opulent cities, but which have, through change, retained some trace of the character derived from their founders. The government regarded these infant colonies with aversion, but could not prevent the population of New England from being largely recruited by stout-hearted and God-fearing men from every part of the old England."

The great powers of that age were unstable and the small nations were very numerous, and both small and great were almost continually at war,

The doctrine we purposed to establish is, that "the future is one of great empires." Great Britain, France, Prussia, Russia, Austria and Italy, on the other side of the ocean, and the United States in America may be said to control the destinies of the other nations.

And never was the doctrine here propounded that small and weak nations are powerless, more fully verified than by the issues of the Berlin Congress of 1878. The small powers were not allowed a voice in its deliberations. Roumania, Servia, Montenegro and Greece were viewed by the great powers as mere geographical expressions, and Turkey has enough to do to live. In a word, the feeble nations must submit to the decrees of the great powers.

The most successful power in Europe in the art of government is Britain. The head of this ancient land of many nations governs upwards of thirty millions of people at home; fifty colonies scattered over the world, and about ninety sovereigns in India, who themselves rule about one hundred and eighty millions of people.

From this brief review of the political consolidations of some of the old states of the world, we purpose to glance at the past and present state of America.

The discovery of America in 1492 marks one of the most important eras in history. By that discovery a vast continent was made known to the old world. Though peopled by millions—though organized and comparatively enlightened nations existed in Mexico, Peru, and other places near the centre of the continent, but little is known of the primitive history of America. And much of its history during the first century after its discovery by Columbus is obscure. Still, enough is known of that period to stamp on America and Europe some of the darkest deeds in their history.

As soon as it was known in Europe that this continent was rich in the production of wheat, corn, rice, tobacco, and other precious fruits; that its forests were immense and contained timber of the most valuable kinds; that its water teemed with all kinds of useful fish; that indigo, chocolate, coffee, quinia, sugar, tar and pitch, were abundant; that the hides of its innumerable wild cattle were valuable; but more especially when it was known that there was gold in Mexico, silver in Peru, diamonds in Brazil, and rich furs in the north, then America became a battle field on which Europeans and the natives of the country became the sanguinary actors.

The continent north and south, east and west, was inhabited by pagan tribes. In Mexico, Peru, and other places near the centre, the native population numbered many millions, and they possessed numerous monuments of civilization; magnificent palaces and temples, highly cultivated fields, gold and silver utensils of the richest kinds, and many other marks of civilization. Mexico and Peru could each look back to a long line of emperors. But all this was doomed to change. In the division of the American continent Great Britain, France, Portugal and Spain became the chief proprietors. · Spain, however, claimed the largest and most wealthy sections.

In 1521, the whole Mexican nation was compelled to submit to the arms of Spain, and in 1532 the last of the emperors of Peru was cruelly murdered in the name of Spain, and his subjects were enslaved. Indeed, the early history of Spain and Portugal in America must forever stand out pre-eminent in the records of human wickedness. They dragged the natives from home and kindred to serve in the mines. It is recorded that eight millions of them perished in the mines of Peru alone. James A. French, A. M., says: "It is the refinement of the Spaniard's cruelty in the settled and conquered provinces excused by no danger, and provoked by no resistance, the details of which witness to the infernal coolness with which it was perpetrated, and the great bearing of the Indians themselves under an oppression which they despaired of resisting, raises the whole history to the rank of a world-wide tragedy, in which the nobler but weaker nature was crushed under a malignant force which was stronger and yet meaner than itself. Gold hunting and lust were the two passions for which the Spaniards cared, and the fate of the Indian women was only more dreadful than that of the men, who were ganged and chained to a labour in the mines which was only to cease with their lives, in a land where but a little before they had lived a free contented people, more innocent of crime than perhaps any people on earth."

Thus did Spain continue for three centuries to sacrifice and degrade the nature of millions of the inhabitants of these regions, in order to replenish her coffers at home with gold, silver and diamonds. The following extract from a letter from Christopher Columbus to King Ferdinand, bears the stamp of the age and country to which the writer belonged,

" Gold is a thing so much the more necessary to your Majesty, because in order to fulfil the ancient prediction, Jerusalem is to be rebuilt by a prince of the Spanish monarchy. Gold is the most excellent of metals. What becomes of those precious stones which are sought for at the extremities of the globe? They are sold and finally converted into gold. With gold we not only do whatsoever we please in this world, but we can even employ it to snatch souls from purgatory and people Paradise."

Shortly after the date of her acquisitions in America, Spain began to decline. The vast stores of wealth obtained from her American possessions had been the means of corrupting all ranks of her people and enervating the spirit of the nation. After a time this source of wealth failed. The chief part of her colonies revolted, and established nationalities for themselves. Mexico, Peru, and other sections became independent; Louisiana and Florida she lost by treaty, and the chief part of her West India Islands by conquest. Out of all her vast possessions in America, Spain holds only Cuba and some adjacent islands, and that by a feeble tenure. Both Spain and Cuba have recently been in the throes of bloody revolutions. In the Imperial State two parties fought for the crown; and Cuba has been fighting to be free from the Spanish yoke.

Thus the political power of Spain ceased on the American main, but not so with regard to her vices. Wherever the kindred tongues of Spain and Portugal prevail, as in Mexico, South America and Cuba, revolutions and vice in their worst forms prevail also. Spain has entailed a dark blot on Mexican civilization. The dregs of Spanish rule, Spanish ignorance and vice continue to curse that country. Mexico for half a century has had no repose from sanguinary revolutions. There seems to be but little hope at present for moral reform. And looking at the vast resources of the country, extent and fertility of domain, mineral wealth, and the genial character of its climate, this is the more to be regretted.

And Spain has been repeating her Mexican history in Cuba. This beautiful and fertile island, situate in a fine climate, and justly named the Queen of the Antilles, contains an area of 48,439 square miles, equal to the aggregate area of Nova Scotia and New Brunswick. In 1870 it contained a million and a half inhabitants, half of whom are white. Of the colored race about four hundred thousand are slaves. The

Cubans have been powerless in the civil matters of the Island. They are heavily taxed in order to enrich the parent State. The revenue is larger than that of the Dominion of Canada. How much longer Spain will be able to govern, or rather misgovern Cuba, it is impossible to foresee. After spending the last five or six years in the most bloody and cruel war, the Cubans have failed to free themselves from Spanish domination.

Spain and Portugal only saw in their colonial revolutions the loss of vast revenues; and, says Humboldt, "the loss of their slaves, the spoliation of the clergy, and the introduction of religious toleration, which they believe to be incompatible with the purity of the established worship."

Portugal held possessions in America second in value only to those of Spain. In 1832 the great Brazilian Empire revolted against Portugal, proclaimed its independence, and conferred the Imperial crown on Dom Pedro, the son of John VI. of Portugal. Thus a monarchy was established on the American continent.

Brazil contains an area of three million square miles. It is nearly fourteen times larger than France. The whole of Britain's Indian Empire would not cover its surface. The soil of Brazil is of unrivalled fertility, and its produce of the soil, mine and sea is varied and rich. Its mountains and valleys are clothed with rich forests. The face of the country is not disfigured by deserts of any great extent. Her rivers, including the great Amazon and La Plata, are large enough to irrigate a continent. Her seaboard, six thousand miles in length, affords great scope for commerce; and, together with her river system, has an ameliorating influence on the colder regions of the country.

Though "discovered by chance," says Southey, "and long left to chance," Brazil is destined to hold an important place in the family of great powers. Governed since her independence by judicious laws, administered by enlightened statesmen, Brazil has far eclipsed Portugal in real progress. This American nation is capable of sustaining hundreds of millions of the human family.

Brazil is the only nation in America whose chief magistrate and his descendants may claim the "right divine" to reign. Whether that right may be considered indefeasible in the future remains to be seen. In North America the people generally believe that the blood of the humblest peasants, who do justice, love mercy, and walk humbly with God, is as royal as that which has coursed the veins of royalty for ages.

The present ruler of Brazil is one of the most enlightened and progressive monarchs of the age; and whether a liberal and limited monarchy, like that of the present, will be continued in Brazil, or whether a republican form of government will take the place of the present, or whether a great monarchical power will prevail in the south, and a great republican power, the United States, in the north, remains for the future to direct.

And France, out of all her vast possessions in North America—embracing the Dominion of Canada and the region lying west of the Mississippi River, now only claims St. Pierre and Miquelon, two insignificant and rocky islands, situate near Newfoundland, and some unimportant islands in the West India group.

French government in America was mild and conciliatory compared to either that of Spain or Portugal. The best example of the progress of French colonization was to be found on the banks of the St. Lawrence. Still, in some respects it was not a fair test. The country near the city of Quebec, where the chief settlements were formed, is far north, and comprised in a narrow valley between the Laurentian and Alleghany mountains. Although this colony was under France for 220 years, and the fertile valley of Ontario was unoccupied, French settlement did not extend. The French people were satisfied to remain where and as they were.

In language, customs, religion, and church and state combinations, Quebec is a type of old France of a century ago. Though adjoining Ontario and New England—though under British rule for more than a century, the progress of assimilation has been remarkably slow.

France and Great Britain fought long and hard with each other in order to obtain and hold possessions in North America. The human life and treasure wasted by these two nations must have been immense. Probably France in the end gained largely by the loss of her American colonies. France is a great power with but few colonies to protect.

Great Britain is the only European nation which has perpetuated its power with any degree of success on the American continent. By conquest and otherwise, Britain claimed the whole Atlantic frontier from the Gulf of St. Lawrence to Florida, including Newfoundland and some West India Islands, and also the French possessions in the north—in all an immense empire. But there is a tide in the affairs of nations. The colonies of Britain, like those of Spain and Portugal, complained of being misgoverned by the parent state. The Imperial government refused redress—a refusal which caused thirteen of her colonies to revolt and erect themselves into a nation. France, pleased to see Great Britain humbled through her own impolicy in America, at once acknowledged the independence of the new Republic as one in the family of nations. Britain, after a series of sanguinary conflicts with the young nation, in which the former was generally the loser, also acknowledged the independence of the United States. Thus Britain lost three hundred and forty thousand square miles of the richest part of her American possessions.

The question might here be asked, why the other British possessions adjoining did not revolt also?

Though a century has passed since the old colonies assumed their independence, the reasons why Quebec did not join the Republic are not without interest to the student of American history. This subject recalls events during which two nationalities were established on this part of the continent; and also the nature of the forces then employed to prevent Canada from becoming a part of the Anglo-American nation, then being established on her southern border.

When the two houses of the British Parliament passed an act, declaring that "the King and Parliament had, and of right ought to have, full power and authority to make laws and statutes of sufficient force to bind the colonies, and His Majesty's subjects in them, in all cases whatsoever;" when stamp duties were imposed on a multitude of articles in the colonies, in order to raise the Imperial revenue at home, the colonies remonstrated by petitions, delegations, and other lawful means.

Against this act of the parent State the inhabitants of Quebec were also opposed. But having fought long and hard, and endured privations of the most serious nature, in order to secure their possessions to France, and failed, it was only natural that the French people would be willing to remain passive, and let the Anglo-Saxons fight against each other. Besides, says the French historian, Garneau, "They ever preserved in their hearts that hatred for the British race, wherever born or located, which they had contracted during long wars; they thus made no distinction in their minds between those of it mingled with themselves, in Canada, and men of kindred blood dwelling beyond, viewing both alike as one body of turbulent and ambitious oppressors."

As time passed their indifference to the sanguinary struggle raging on their southern border gradually subsided, and they strongly sympathized with the other colonists in their opposition to the Imperial demands.

The new republic, however, in its declaration of the rights of man, gave great offence to the Roman Catholic clergy of Quebec. In the declaration Britain was blamed for setting " up civil and spiritual tyranny in Canada, to the great danger of the neighboring Provinces."

Notwithstanding this strong declaration against the civil and religious institutions of Quebec, the sympathy of the French laity continued with the other colonists, and when France acknowledged the independence of the republic their sympathies became still stronger in favor of the New England States.

At the conquest, Great Britain secured to the French colony on the St. Lawrence their religion, language, laws, customs and other institutions. These were held sacred, especially by their clergy, who, when solicited to join the Republic, replied: " The British government has left us nothing to wish for. All our monasteries are now in full possession of their own; our missions are in a flourishing state," and " the military authorities are ordered to do honor to our religious outdoor ceremonies." Believing that their peculiar institutions would be protected and better respected by Britain than by the republic, "the Bishop of Quebec addressed an encyclical letter to his flock, exhorting the faithful to be true to British allegiance, and to repel the American invaders. He strove to prove at the time that their religion would not be respected by Puritans and independents if these obtained the mastery in the struggle going on, and that it would be folly to join them." "These sentiments," continues Garneau, "were more widely developed by him afterwards in a lengthy pastoral letter, published next year. Meantime neither the proclamation nor the encyclical were able to move the Canadians from their state of apathy." Notwithstanding the sympathies of

20

the laity with the revolutionists, the "clergy and seignors re-
solved to resist every assault of the Anglo-Americans, and retain their country for monarchical Britain, 3,000 miles distant,
a patroness all the less likely for that remoteness to become
perilously inimical to Canadian institutions." However,
through the influence of the clergy, the Canadian people became "gradually cooling in their republican tendencies, and
continuously influenced by the calming and efficacious advice
of the clergy and burgesses—the latter of monarchic sentiment," the laity changed their policy and fought against the
republic; and, says the historian, "it may be fairly assumed,
then, that to the clergy of Canada at this juncture was Britain
indebted for the conservation of the dependency."

Thus Canada was to remain monarchical, and the French
people a cold abstraction. Had Quebec joined the United
States in 1775, Canada would have obtained the rights and
liberties which she rebelled in order to obtain in 1837;
and Great Britain would have been preserved from American
complications, which have caused her much trouble, and
from wars and war debts of immense magnitude

In the political division of North America, Great Britian
has been singularly unfortunate. Out of the wreck of her
valuable colonies only a few isolated patches of habitable
country remained to her lot, and even what remained has not
all been retained. It would transcend our limits to give a
full historical retrospect of the boundary and other disputes
which have arisen between Britain and the United States;
suffice it to say, that the territory which history has cast to
the lot of the Dominion of Canada is not the old French
empire unimpaired. Many millions of acres of lands claimed
by the Provinces, besides other valuable resources, have been
given by Great Britain to the United States against the pro-
tests of the Provinces. The residue is totally unlike any
other country in the world.

After the revolt of the "Old Colonies," Great Britain
made no effort to extend or strengthen the southern frontier
of her possessions; but on the contrary, in her boundary disputes she gave piece after piece to the Republic, amounting
in the whole to forty or fifty millions of acres. Indeed, seeing the inferior character of her remaining possessions in
North America, it is doubtful if her leading statesmen felt a
strong desire to retain them.

Great Britain allowed the great empire of Louisiana, on her southern border, to be ceded and receded down to the beginning of the present century without making an effort to claim any part of it, France claimed this region by right of discovery in 1618. It was ceded to Spain, and receded in 1800, and sold by France to the United States in 1803, for the insignificant sum of fifteen millions of dollars. Out of this region the Republic has erected Arkansas, Missouri, Wisconsin, Iowa, and the territories lying west of these States.

The United States continued to add largely to her national domain. In 1819, in the settlement of her disputes with Spain, the latter ceded Florida to the Republic. About the middle of the present century she obtained by conquest and purchase from Mexico, the present States of Texas, Nevada, California, and the Utah, New Mexico and Arizona territories; and in 1867 she purchased Alaska from Russia for about seven millions of dollars.

Thus the Republic comprises a great cluster of States, the greater part of which is inaccessible to a menacing and destructive invasion. If such growing power existed in Europe it would be viewed with great suspicion by neighboring nations, but being in America no nation interferes. The only recent instance of European interference in the affairs of America was the effort made by France to establish a branch of the Royal family of Austria on the throne of Mexico. But the effort failed in the most disastrous and melancholy manner; and Mexico is still feasting on bloody revolutions, and will probably continue to do so until annexed to the United States.

It was long believed that the slavery question would be the means of rending the United States; that at the close of the rebellion the union would be disunited, and that several powers would spring out of the wreck of states, when a Canadian nationality might have a fair chance to live and breathe more freely. After the rebellion was suppressed, and the States set free, and the nation reconstructed on a more permanent foundation, public opinion changed. It changed also in regard to the relations of British North America to the empire and to the United States. Immediately after the close of the civil war in the States, Great Britain, by means of treaties, got her chief disputes with the Union settled, and then made a safe retreat from North America, taking her soldiers and munitions of war with her. Before leaving Canada to the Canadians she gave the United States the free navigation of the River St. Lawrence forever; also the command of the San Juan navigation, and the use of the Canadian fisheries. But the United States government refused to refund to Canada the cost of the Fenian raids, which amounted to a million and a half dollars.

As already stated, the policy of the world is the unification of feeble nations into great powers. By this law of development the treaty alliances guaranteeing the integrity of small powers have been disregarded; and the small and feeble states of the world are being encompassed by great powers; and thus a more enlarged concentration of power and authority is placed in the hands of a few great powers, instead of being divided among a host of feeble nations.

With the exception of Great Britain but few nations have colonial possessions. Indeed, there is not the same desire as formerly to acquire distant regions. And the opinion is fast gaining ground in Britain that her colonies, especially those of North America, would be a source of weakness to the nation in the event of a general war.

The Anglo-Saxon mind of the British Islands, whether at home or abroad, whether under monarchical or republican institutions, is peculiarly adapted to govern large masses of the human family, and according to diversified requirements. In North America a century ago, in opposition to the "right divine of kings," this race laid deep and broad the foundations of a great Republic; indeed, a greater Britain than Britain itself, and free from the mediæval element which shaped the structure of European society—church and state, monarchs, popes and aristocracy, with free and unsectarian schools, and an open Bible as the key to the structure. And in British North America the same mind under different circumstances has made great progress, indeed a remarkable progress when the two obstacles are considered. While the States governed themselves the Provinces were governed by a power which is three thousand miles off, and which for a long time claimed a monopoly in colonial trade. Besides, the progress of Canada proper was retarded by intestine troubles, arising out of conflicting races and interests.

Wherever groups of Anglo-Saxon communities exist, political union of some kind generally takes place. It was so in the British Islands, in the old thirteen colonies of Britain, and in British North America. The Australian colonies also are in favor of union.

But there is no part of the world in which the Anglo-Saxon mind prevails, where their separation is so unnatural and inconvenient as it is in the United States and Canadian Provinces. Geographically these two countries belong, so to speak, to each other. The habitable parts of Canada are separated from each other by immense regions of infertile country. The settlements of the Dominion east of Lake Superior are connected to those of the States in front by roads, railroads, telegraph lines, canals, rivers and lakes, while the only connection between the Upper and Lower Provinces is a long stretch of worthless lands, through which a railroad passes. The vital and varied interests of the States and Provinces are so blended, and their natural, social and moral ties are so closely united, that the political union of these two branches of the same family cannot be long delayed. Indeed, in conformity with the antecedents of the Anglo-Saxon race, and in accordance with the spirit of the age, a strong union sentiment has prevailed on both sides of the international boundary. And, as events unfold themselves and the true position and relations of the habitable parts of the Dominion become better known, the united destiny of the States and Provinces become more and more manifest.

No matter what policy the Dominion may pursue, her scattered settlements are confronted on all sides by almost insurmountable barriers. Her natural markets, as we have shown, are not within her own boundaries nor in Great Britain, but in the United States. Large numbers of the population of the Dominion are continually emigrating to the States, thus binding the kindred ties of the two nationalties still closer as time passes, and thus a permanent confederation of minds is being established.

In the union of 1867 each Province sacrificed some long cherished notions with a view to the general advantage of the whole; so it is in all voluntary unions, whether of states or of churches. In Scotland and Ireland, at the time of their unions with England, many were opposed to the abolition of their local parliaments. That objection would not arise against a union of these States and Provinces, as each would retain its local legislature. And if the Dominion Parliament,

158

with all its costly machinery was abolished, no one need,
and probably very few, would be sorry.

Obliterate the unnatural boundary which divides the Provinces from the States --remove the custom houses from the
boundary line, and give these two countries free and unfettered trade with each other; give the Dominion unrestricted
trade with the forty-five millions of progressive people on the
front, and both countries would soon have proof positive as
to where their true interests lie. The kindred and geographical ties, and the commercial relations which bind the British
Islands so closely together to-day, are not more binding than
those which nature has established between these American
branches of the same family. Indeed, no one has yet been
able to show how either Canada, Britain or the States can be
benefited by the political separation of these American
nationalities. There may be some whose monarchical predilections lead them to see in vision the rise of a monarchy
among the mountains of Canada, under the ægis of which
they may hope to be the recipients of greater honors and
emoluments of office than under republican institutions.

Having viewed the position and relations of these adjoining nationalities with all the care and impartiality we are
able to command, we are unable to see how it is possible,
under existing circumstances, for the Dominion to develop
its own resources unless in a commercial union with the
United States. And how, in the event of war, the Dominion
can be defended against the United States in force is more
than we can divine. It is however to be hoped that before
war come, able advocates of the union of those two countries will rise to the dignity and importance of the subject,
and show throughout the length and breadth of these States
and Provinces, that the union of the whole is a necessity.
Then the dangers which confront the Dominion will vanish,
and the happiness of leaving to our children the heritage of
an honorable citizenship in a united and prosperous nation
will be a great reward.

## INDEPENDENCE OF CANADA.

There is much in the early history of America which reminds us of what the historian Robertson says of the beginning of most all commonwealths:—"Nations as well as men arrive at maturity by degrees, and the events which happened during their infancy or early youth cannot be recollected and deserve not to be remembered."

It is only as lights, casting their shadows upon the future, that many of the events in the history of this country deserve to be remembered. The nations of Europe holding colonies on this continent claimed to have a divine and indefeasible right to exercise regal power over them. And the power was generally used with a strong arm. "The doctrine," said Lord Macaulay, "that the parent State has supreme power over the colonies is not only borne out by authority and precedent, but will appear when examined to be in entire accordance with justice and with policy. During the feeble infancy of colonies, independence would be pernicious or rather fatal to them. ....There cannot really be more than one supreme power in a society....There ought to be complete incorporation, if incorporation be possible. If not, there ought to be complete separation. Very few propositions in politics can be so perfectly demonstrated as this, that parliamentary government cannot be carried on by two really equal and independent parliaments in one empire."

After the revolt of the "old thirteen colonies," one of the most perplexing questions of a colonial nature Great Britain had was, how to govern and protect the fragments of habitable country that remained to her lot in North America.

During the early history of this country Great Britain held a monopoly of the trade of her colonies, and this monopoly, says Adam Smith in his *Wealth of Nations,* was "the principal, or more properly perhaps the sole end and purpose of the dominion which Great Britain assumed over her colonies."

At the same time the colonies had the privilege of underselling foreign countries in the British markets at home.

Long ago this was changed. By degrees Great Britain has relaxed her control of the legislation of the colonies. The latter buy and sell where they please. And the commercial relations between Britain and her colonies is on a par with that between either of them and foreign countries, except that the trade system between the colonies and foreign countries is subject to changing legislation, while that between the parent state and foreign nations is generally based on treaty obligations. Consequently Britain's commercial advantages arising out of the colonial relations are but few. When the monopoly ceased her colonial policy may be said to have changed, and her control of the colonies became more and more relaxed, especially with regard to British North America. Her avowed policy is to let the Dominion of Canada shape her own destinies as she may think best.

It cannot be doubted that the power and policy of the United States had much to do in giving force to the present colonial policy of Great Britain. Her almost ceaseless troubles with that power, arising out of her political connection with these Provinces, has given rise to a strong feeling in the British isles to get rid of these Provinces altogether. In the last century and early part of the present, Great Britain was engaged in many of the wars which devastated large sections of the world. She made vast territorial acquisitions, which were then considered conducive to the power and prestige of the nation. The colonies now make their own laws, and trade where they please without giving the mother country any preference. Hence, the chief voice in the British isles is in favor of a separation of the colonies, or the greater part of them, from Great Britain. It is asserted that the colonies yield no tribute to the support of the nation; they cost much, give great trouble, and in the event of a general war, they would be mere hostages to the enemy.

There is another and not less important point connected with the colonial question. That is, as the number of great powers increase in Europe, the necessity increases for each power to concentrate its forces at home. France, Prussia, Russia, Austria, Turkey and Italy have but few, and some of them no distant possessions. Each of these nations keep immense forces within the home domain, and are always ready for war, and always watching each other; and Great Britain has enough to do in watching them all, and keeping the road open to India, without being burdened with the Government and protection of colonies which render no service to the empire, but are a source of danger in the event of war. Hence in the Crimean and Abyssinian wars Great Britain made no territorial acquisitions; and she ceded the Ionian Islands to Greece, and Canada to the Canadians. Indeed, any territorial acquisitions she has recently made have been caused more by the exigencies of the occasion than from any apparent desire to acquire more colonies.

In pursuance of her colonial policy Great Britain has, step by step, made these Provinces more free, and now virtually *independent.*

In 1867 the chief Provinces of British North America framed a Constitution for themselves; they fixed the limits of their own jurisdiction, which the British Commons, Lords and Queen with great pleasure stamped with the force of law. Hence Canada has started with a written constitution, under which she has assumed national responsibilities. This constitution is not legislative like that which binds the British Islands politically together; but as a preparatory step to Canada's manifest destiny. We live under a Federal constitution, framed, as near as monarchical principles admit, after that of the United States.

There was, however, a two-fold interest involved in the union question. Great Britain urged the unification of the Provinces as the only feasible means of relieving her from further liabilities in regard to their defence; and the two Canadas urged the union as the only means of securing the adjustment of their sectional and party differences—differences which had for many years made government almost impossible. However, in uniting the Provinces adopted the course which impels weak communities throughout the world to pursue.

Nearly all the feeble States whose integrity Great Britain guaranteed have disappeared as such from the map of Europe, and are now incorporated with great powers. Indeed it is almost impossible to secure respect for the frontiers of weak and isolated communities, especially when such communities are contiguous to a great power.

How long the Dominion may remain a safe resting place for the Canadian people we cannot predict. Certainly it is not a convenient one. The only political tie which connects Canada

21

with the Empire is the Governor-General, and that tie, in consequence of its vice-regal pageantry, which renders it unsuitable for American society, and its great cost to the Canadian people, is being weakened and may soon be severed. 'Since the union of the Provinces nearly all the old ties which connected each Province with the Empire have been dissolved. In fact a radical change has passed over the whole system of colonial government in North America. Canadians may well ask, what about the future of the Dominion?

The following citations and deductions from the public speeches of leading British and Colonial statesmen, and from the leading press of the British Isles show that Britain's colonial policy, especially as regards British North America, has been undergoing remarkable changes.

As far back as 1828 Mr. Huskisson, then Colonial Secretary, said: "He thought the time had come for the separation of Canada from the mother country, and her assumption of an independent state." And in the debate on the ordnance estimates he said: "If he could be positive that the amount of the present vote was to be expended with the positive certainty that in fifty years to come the Canadas were to be free and independent, he would not hesitate, but would as heartily give his vote under such circumstances as he would give it now." The fifty years are now expired, and Canada is virtually free.

On the same occasion Lord Howick said: "There could be no doubt that in time all our foreign colonies would become independent of the mother country. Such an event was certain, and we ought in time to prepare for the separation, not by fortifying the Canadas, but by preparing them to become independent."

Twenty years after this policy was announced Mr. Richard Cobden, in a letter to Mr. Sumner of the United States, said: "I can assure you that there will be no repetition of the policy of 1776 on our part to prevent our North American colonies from pursuing their interests in their own way." The Earl of Ellenborough, in 1854, said in the House of Lords: "He hoped the Government would communicate with the North American colonies with the view to separation." On the latter occasion Lord Brougham said: "He was one of those who desired a separation of Canada from the mother country. The idea," he said, "was not novel; it had been entertained and pressed by many eminent men, It was an opinion shared in by Lord Ashburton and Lord St. Vincent." A member in the House of Commons on a recent occasion declared "that the relation between Canada and Britain was rotten and mutually deceptive;" while another, a Cabinet Minister, said: "He looked forward without apprehension and without regret to the separation of Canada from England." . In 1864 Lord Derby, a former leader of the great Conservative party in England, said: "In British North America there is a strong movement in progress in favor of federation, or rather union of some shape....We know that these countries must before long be independent States."

And nearer to the present time we find the foregoing policy fully confirmed by many of the leading statesmen of the day. Mr. W. E. Gladstone, one of England's great statesmen—one, too, whose influence was fully exerted, and indeed was powerful in saving the British nation in 1877 and 1878 from being involved in a war with Russia—when leader of the British Government in 1870, in advocating the separation of these colonies from the Empire, said that "the present Government do not claim the credit of adopting or introducing any new policy,....persons of authority of every shade of politics have adopted it. When," as he said, "we see a country like the United States, that sign of immense human energy, extending itself continuously over that vast continent," Mr. Gladstone was no doubt convinced that the tie which connects these colonies to the Empire is but feeble. And during the discussion in Parliament on the Imperial guarantee in aid of the construction of the Intercolonial railroad, said he "should find it impossible to justify" the guarantee, "except on the conditions that England would be free from responsibilities as to the defence of Canada....It cannot be too distinctly stated that the defence of the British North American colonies is a very heavy charge indeed, and it is our duty in every way to get rid of it....It is in this view that we look upon the plan of uniting them." And again, in the debate on the bill guaranteeing a Canadian loan of five and a half millions of dollars in aid of the erection of fortresses in the Dominion, he said: "This guarantee was part of the price England paid for being relieved of the obligation to protect Canada by military force. England had now arrived at the state of things in which Canada was to undertake almost entirely its own defence;....

in which England would be relieved from all demands upon her exchequer on account of Canada."

Mr. Lowe, in a recent speech in Parliament, said we should represent to Canada "that it is perfectly open to her to establish herself as an independent republic; it is our duty, too, to represent to her that if after well-weighed consideration she thinks it more to her interest to join the great American republic itself, it is the duty of Canada to deliberate for her own interests and happiness.

As soon as the chief Provinces united Great Britain withdrew her troops and military stores from the Dominion of Canada. This is the first time in her history of abnegating her right to occupy and hold fortifications which she had constructed at great cost. But times have changed; that "sign of immense human energy" has extended itself continuously across the American continent, and consequently the British people at home have concluded to let Canada shape her own destiny. On the withdrawal of the troops Earl Granville said: "I quite admit that the general tendency of our policy is to leave to a country like Canada....the duty of self-defence."

It would require superhuman bravery indeed for four millions of people, scattered here and there along thousands of miles of a defenceless frontier, to defend themselves against more than ten times that number. "The practical effect would be," as Earl Carnarvon said, "that the North American colonies would be left to the mercy of any set of privateers in the case of a war breaking out with the United States."

In the House of Lords in 1870 Lord Grey said that "the principles laid down by successive Colonial Secretaries must necessarily lead to a dissolution of our Colonial Empire." In the same debate Lord Northbrook pointed out "the necessity of a greater concentration of troops at home for the security of the Empire."

Lord Russell, who had a strong desire to see a closer union of the colonies with the Empire, frequently expressed doubt as to the future of Canada. He said: "If it should ever be their wish to separate from this country, we may be ready to listen to their requests and to concede to their wishes in any way they may choose." Or "if the North American Colonies felt themselves able to stand alone, and showed their anxiety either to form themselves into an independent country, or even to amalgamate with the United States, he did not think it would be wise to resist that desire." Lord Normanby said: " England desired no pecuniary benefits from the colonies."

The Hon. Joseph Howe, when in England, during the discussion on the union of the Provinces, said he heard a noble Marquis say that "those British Americans may go and set up for themselves when they please;" and further, "that we might annex ourselves to the United States if we pleased, and no power would be used to prevent us." He said: "The Marquis made the statement, and not a man rose to contradict him."

But few, however, had greater opportunities than Lord Monck of knowing the position, resources and defences of Canada. He was Governor General of British North America during the protracted discussion on the union of the Provinces, and was the first Governor General of the Dominion of Canada. After his return to his place in the House of Lords he said: " It is in the interests of the mother country that" these Provinces " should be taught to look forward to independence....He believed the policy of Government tended towards such independence, and it was on that account he gave the Government his support." He alleged that the tie which connects Canada with Britain was a mere sentimental one, that the connection had ceased its uses, and that the colonial relations to Britain were dissolved when Confederation was consummated, and that the true mission of Canada was to proclaim its independence.

And Sir George Campbell, formerly Governor General of India, said: " Canada has grown to maturity.....I would let it go free without more delay, and would relieve this country of the many embarrassments to which the connection may give rise. Canada I believe to be, under present arrangements, a burden and a risk to us."

In his work, *Greater Britain*, p. 379-380, Sir C. W. Dilke says: Canada " draws from us some three millions (sterling) annually for her defence; she makes no contribution toward the cost, she relies mainly on us to defend a frontier of 4,000 miles, and she excludes our goods by prohibitive duties at her ports. In short, colonial expenses which, rightly or wrongly, our fathers bore (and that not ungrudgingly) when they enjoyed a monopoly of colonial trade, are borne by us in the face of colonial prohibition. What the true cost to us of Canada may be is unfortunately an open question, and the loss by weakening of our home forces we have no means of computing. ...We cannot but admit that we pay at least three millions a year for the hatred that the Canadians profess to bear toward the United States."

Since the union of the Provinces England's cost for their defence is comparatively small, but in the event of war with the United States would be immense.

Among the advocates of a closer union of the colonies with the mother country the Right Honorable W. E. Forster is foremost. In a public speech in 1875 he expressed a hope that some practicable plan might be devised, but could not then name a plan. He said "that the common belief was that the colonies must some day become independent, and this common idea would, he feared, become one of those which realize themselves." After combating the arguments of separationists, he said the arguments from the difficulties of colonial connection in time of war, however, he regarded as much more important and less easy to meet. It applied only, however, to Canada. Her disputes with the United States might, he admitted, involve Britain in war, and her border line would evidently be a source of great weakness during hostilities with the great Republic.

"Many plans," said Lord Derby in 1876, "have been proposed for connecting Australia and Canada more closely with this country, but never yet one that looked as if it would work." Lord Russell at one time proposed that the British Parliament should guarantee the security of the colonies, and that the latter should buy British protection by contributing three or four millions sterling to the army and navy estimates of Britain, and undertake not to charge more than ten per cent. *ad valorem* on British produce. Evidently the colonial system has passed the stage of growth to which such a scheme would have been appropriate. And the scheme of colonial representation in the Imperial Parliament was pronounced by Mr. Gladstone as "altogether visionary. We cannot," he said, "overlook the countless miles of ocean rolling between them and us." And the press has frequently reiterated the same views as to the future of these Provinces.

The London *Times*, the great exponent of public opinion in the British Islands, has repeatedly called on Canada to assume her independence or join the United States. In a review of Lord Monck's speech made at the opening of the first Parliament

of the Dominion of Canada, in 1867, the *Times* said:
" Empire never spoke with so small and still a voice as when
England humbly suggested and greatly aided the idea of a
Canadian Confederation. She could say little for it, except
that there the colonies were, they had a common origin and
allegiance, they had common dangers, and inasmuch as they
had diversity of conditions and interests, they had matters to
settle. There was a sort of moral unity; it had better be
made political....Lord Monck trusts and believes that the new
nationality will extend from the Atlantic to the Pacific, but
political faith overreaches itself in a conception so vast and so
loose, in frontiers so extensive, and in conditions so infinitely
various. It supposes a nationality able to command the two
oceans it touches, and to raise a barrier of law and moral force
extending near three thousand miles between itself and the
most powerful and aggressive state in the new world. · We
look in vain for the body, the. vital organs, the circulation,
and the muscular force that are to give adequate power to
those wide-spread limbs....However, there it is; and, as we
say both at the first and at the last scene of human existence,
where there is life there is hope. There is a strength in weakness....The weakness of the immense frontier is confessed by
those who ask defences, and proclaimed by those who think
all defences vain, and have no wish they should be otherwise."

And the occasion of Lord Dufferin's departure for Canada
led the London *Times* to remark: "It is useful to call to mind
now and then the strange and anomalous character of institutions of which Lord Dufferin is the embodiment. A dependency which in no way depends; an administration conducted
in the name of the British Crown and under the authority of
the British Parliament, but yet controlled in everything, great
and small, by the votes of the Provincials; a liability of the
Dominion to be engaged in war and subjected to invasion for any
dispute of British origin, even though it may have begun in
China or on the shores of the Mediterranean; a liability of
Great Britain, on the other hand, to defend a frontier of many
hundreds of miles, on which she has not now a single soldier,
and to which, during a long part of the year, she could not
send reinforcements; an integral part of an Empire from which
the same Empire derives neither financial support, nor military
strength, nor diplomatic influence, but by which it is, on the
contrary, sensibly weakened and hampered.— Such are the

phenomena which the British authority over the Dominion presents. We cannot but feel that a state of things so out of accord with any theory of government, and so apparently unsuitable, requires very delicate treatment. The question is whether a statesman should seek to divine and anticipate the course of events, or whether he should trust that the Dominion will, by some spontaneous action, work out its own destiny in the fulness of time. The latter seems to be the policy which recommends itself to British statesmen."

Again the *Times*, referring to the heavy taxation which the vast armaments of Europe impose, said: "The new world is now to all expectations free from the war feud. One government possesses a preponderant influence over the Northern Continent, and will probably within our time absorb the feeble Mexican Republic....The Canadian Dominion may or may not have a similar destiny, but in any case the people who occupy it need never feel the burdens of a military system."

The *Examiner*, in its review of the San Juan award, said: "The North American Federation is quickly preparing itself to be an independent nation, and, after that, if it joins the older federation of the United States English statesmen need not be surprised, and perhaps no one need be sorry." The *Westminister Review* said: "Canada has been deliberately and officially warned that the days of British connection are numbered, and that she must determine on a new state of existence." And the *Edinburgh Review* said, that the separation of Canada from the Empire " is not a thing of yesterday, and is not due entirely to the action of the Ministers of the day;" and describes Canadians as " retainers who will neither give nor accept notice to quit."

And in Canada we find the leading statesmen and press preparing the public mind for the inevitable crisis which is step by step closing on the Dominion.

Mr. Mackenzie, the late premier of Canada, in a public speech in Sarnia in 1875, said: " We have long ago passed the bounds of an ordinary colony of Great Britain; we have assumed the proportions of a nation."

In regard to treaties, he said: "I have no idea that any British statesman will think after this of interfering....to procure treaties with the United States or any other power we are dealing with when a treaty is to be made purely in Canadian interests,...The responsibility of doing it will be technically with the Imperial Government, but with the Canadian Government will rest the responsibility to the people of Canada, and the management of such affairs through Imperial officers."

Thus it appears British Ministers at foreign courts are simply agents of the Dominion when Canadian interests are the question.

Mr. Mackenzie, with whom faith indeed is the evidence of things unseen, further says, that "it is decreed as inevitable that there shall be at least two systems of political government upon this continent."

And Lord Dufferin saw in vision the rise of a Canadian nationality. When in England in 1875, he asserted that Canada believes she "is destined to move within her own separate and individual orbit;" that Canadians "have exuberant confidence in their ability to shape their own destinies to their appointed issues;" and that "they desire to claim their part in the future fortunes of the British Empire, and to sustain all the obligations such a position may imply."

The Hon. Mr. Blake, a member of the last Dominion Government, in a public speech in 1877, made some pointed remarks with regard to Canada's present and future relations with the Empire.

Believing that the Dominion of Canada is "quite competent to determine what laws should regulate our maritime concerns, and to interpret and administer the laws we make, without resorting to the British Parliament for legislation," and believing that the Governor General's commission and instructions from the Imperial Government "are no longer suited to our circumstances," Mr. Blake was charged to discuss these and other subjects, and if possible obtain a "fuller measure of self-government." The British Government readily acceded to his wishes.

It appears that a special code of maritime laws has prevailed for many years on the United States side of the great lakes, by which that power could take "a lien on the ship in respect of certain classes of wrongs and contracts." Canada has since inaugurated a system of maritime laws with regard to the lakes, and probably will before long extend her "legislation to the maritime concerns of the sea-board."

The Governor General's commission and instructions were to be recast in a Canadian mould.

22

With regard to the future of Canada, Mr. Blake, in the speech referred to, said: "The present form of connection is not destined to be perpetual. My opinion is, that the day must come when we shall cease to be dependents." In the event of such a change, he preferred some kind of partnership with Great Britain in which the Dominion would be bound to pay her share of the costs of maintaining the Empire.

To one who has carefully studied the conditions and relations of the Dominion, it must require a wonderful stretch of imagination to see in the future of Canada a national foundation. In no part of the world claiming national existence are there to be found so many anomalous conditions united in one country. Here is a confederation stretching from ocean to ocean at the widest part of the continent, without the least possibility of having a territorial consolidation of settlements. The settlements in the Dominion are scattered here and there, and far apart, along side of one of the most powerful and aggressive nations in the world.

And to rest our hopes of protection on a power which is 3,000 miles from our shores, even suppose such guarantee could be obtained and maintained against the United States, which in a short time will possess double the population of the British Isles and Canada together, shows a marvellous amount of faith.

It is obvious that Canada is at liberty to raise an army, build a navy, and erect fortifications, borrow money, or unite with the States, or do none of these things. Whatever shape the future of the Dominion may assume, it will be American and not European in its tendencies.

The inhabitants of the States and Provinces are drawing closer and closer to each other in all that pertains to the social relations of life. Their educational, currency and postal systems, have been assimilated. Even the railroad gauge of the Dominion has been changed to suit that of the States; and their legislative systems have been assimilated in many respects. Each country is a confederation, the one of States, the other of Provinces.

Still, so long as there is an international boundary between these two countries, criminals will cross it and escape justice; debtors will cross it, and creditors will suffer; and the revenues of both countries will suffer by illicit trade, notwithstanding

two costly lines of custom houses. Indeed the ways are
many in which annoyances arise.

In fact the disfiguration of the Canadian Dominion will
always give_rise to trouble between these nationalities in
North America. There is no. political division 'n the world
where a union is so absolutely necessary as between these two
countries. Though unions are the order of the age, the union
of the Provinces, arising out of the peculiar situation of the
country, has been of comparatively little service to the
Dominion. Since the union the Dominion has been building
upon a false capital. A balance of trade amounting to two
hundred and thirty millions of dollars has been incurred
against the country; the public debt of the Dominion has been
doubled since Confederation, and is in a fair way to be
doubled again before long; and the public revenue has been on
the decline for some years.

At the time of the union many in the Provinces preferred a
union with the United States, but the time was unfavorable.
The Republic had just come out of great tribulation. Her
commerce was prostrate, and her political machinery out
of joint. In a word, a great crisis had transpired, during which the United States incurred an immense debt, and
a large public debt at that time was viewed by the Provinces
as a great calamity. Hence the feeling in the Provinces in
favor of union with the States was not half as strong as it
would have been under more favorable circumstances. The
last ten years, however, have produced great changes. The
United States is now a great consolidated power, slavery is
abolished, peace restored, and the public debt greatly reduced.

That a higher destiny awaits the Canadian Dominion than
that of being a petty nation, without a possible consolidation,
without national advantages, we do not doubt. That the
union of these two countries would place both on the high
road to a mighty destiny, which would far eclipse that of any
other country in the world, is obvious.

## CANADIAN DEFENCES.

The defence of these Provinces has long been a subject of deep interest to Great Britain as well as to the Provinces themselves. And in recent years as time passed the interest felt became more intense. The Provinces began, as it were, to see how one Province looked in the light of another; and in what relations they stood to the mother country and the United States. And after the close of the rebellion in the States, when the latter became consolidated, Great Britain urged the Provinces to unite; and when united, she withdrew her military forces, guns and stores.

Previous to the union Great Britain maintained strong garrisons at Quebec, Halifax and Kingston; and small forces at each of the other chief cities of the Provinces; at what cost we have no means of knowing. It must have been great. Previous to that date, the despatches from England urging these colonies to prepare for war would fill a respectable volume. The latter declined, or did not feel able to do more than organize a few battalions of militia. Indeed, their chief dependence for protection was upon Great Britain.

The withdrawal of the troops, together with the cessation in the construction of fortifications by Great Britain, has, in a pecuniary point of view, been a great loss to these Provinces; while, no doubt, it has been a gain to the tax payers of the British Islands.

Great Britain has ceased pleading with these Provinces as formerly to prepare for war, and the latter are fast losing sight of the question of defence. Indeed, with the exception of some small expenditures by the Dominion, but little attention has been paid to the question. The fortifications, especially those of "Quebec City and District," says the Minister of Militia, "are rapidly going to destruction from want of attention and repair."

However, when we consider the peculiar formation and relative position of the chief settlements of the Dominion, the prejudices and institutions of the people, the indefensible character of the Provinces, the intercourse between the latter and the United States, the facilities for transit between these two countries, and the absolute necessity for such intercourse, and above all, the kindred relations of these two nationalities, it is not probable that the Canadian Dominion will ever become a military and naval power, and it is doubtful if the resources of the Provinces could be made fully available in the event of war with the United States. There are but few families in the Provinces but what have kindred in the States, while there are many millions in the States who have no kindred in the Provinces; hence the latter are not so fully bound by ties of kindred to abstain from hostilities against the Dominion.

Whether Canadians are willing or not to entertain the question of defences, we cannot in justice to ourselves ignore it. We are shut up to facts. Indeed, if there is any one thing connected with the future of this country more clearly defined than another, it is this: The Dominion cannot defend itself against the United States in force. Then follows the important question: Will Great Britain attempt to defend it? In the answer to this question every Canadian has a deep interest.

Great Britain is unquestionably a great power in Europe. She has long occupied a highly important position in the family of European nations. Her territorial isolation at home, her powerful navy, vast commerce and enlightened statesmanship, have aided in securing for her great power and influence in Europe and other parts of the world. In North America, however, her power has been, in a measure, impaired. Out of her most valuable possessions of a century ago, a hostile power has taken root, and since that time spread itself far into Mexico on the south, west to the Pacific Ocean; and north, by continual abstractions from British territory, nearly as far into the arctic regions as the continent is habitable.

A century ago, four millions of people, organized into a Republic, began creating a nation. Now forty-five millions are giving dignity and force to national life. The situation and vast extent of the United States in the most favorable part of the continent; her enormous stores of natural and developed wealth—her extensive seaboard and inland navigation, and the rapid increase of her population, representing

all nationalities, are combined in developing the most extensive agricultural, commercial and manufacturing industries
to be found in the world. And the amazing prosperity which
has attended her policy eloquently points to a great and prosperous future. Indeed, no country having so many discord-
ant elements in her population has done more in so short a
time to secure the civil liberty and education of her people
than the United States. The increase and diffusion of wealth,
free and unsectarian schools, absolute freedom of religious
worship, without an established church, and above all an open
Bible, has given great effect to the moral sentiment and energy
of the nation. At the present rate of increase, the United
States may, before another generation of our able men pass
away, contain one hundred millions of people. Already she
holds a proud position among the great powers of the world.
Thus situated and endowed, the United States will always
command great influence both in Europe and America. In a
war with Mexico or Canada, or even with both together, she
cannot fail to be a great power. And no nation has learned
better how to make war pay. It was only in the recent rebellion that she suffered to any great extent. That was a struggle,
however, on which great issues depended; the suppression of
slavery, the unity of the Republic, and the shaping of the
destinies of North America.

That the United States is desirous of extending her frontiers
to the outskirts of North America there is but little doubt;
and in the event of her doing so will any European nation
interfere with her aspirations? The United States policy, as
enunciated, is not to interfere in European complications,
while she considers any attempt on the part of European
Governments " to extend their system to any portion of this
hemisphere as dangerous to her peace and safety."

However, the desire on the part of the United States to extend her boundaries may not be the immediate cause of war.
The peculiar configuration of Canada will always tend to com-
plicate her movements, and make her dependent upon the
Republic for vitality and even existence as a separate
nationality.

The habitable parts of the Dominion are scattered far apart
along a frontier of nearly four thousand miles. Between
the chief settlements there are immense areas of infertile lands,
and the total population is comparatively small and utterly
unprotected.

However, no settlements in the world are so completely protected as are those of the Dominion, on the north, which lie between the Gulf of St. Lawrence and the Pacific Ocean. The untrodden solitudes which comprise the arctic slope of this continent are a complete defence against invasion. Nowhere does desolation reign so completely. This is the battle-field which the opposing forces of heat and cold have selected. But the south side of the Dominion, as Lord Derby said, "is peculiarly open to aggression."

The settled sections of Ontario and Quebec lie in a long narrow strip, completely protected on the north by the laurentide mountains. On the south they are entirely unprotected.

Near the lower part of the Province of Quebec the United States, for a distance of sixty miles, is within from twenty-five to thirty miles of the river St. Lawrence. Hence the hour-glass shape of the geographical tie which connects Quebec to New Brunswick will render the Intercolonial railroad peculiarly open to destruction in the event of war with the United States.

This railroad was built chiefly for military purposes, but its proximity to the international boundary will render it almost useless as a means of defence.

The frontier line of New Brunswick, Quebec and Ontario for a distance of about 1,500 miles is undefended, and we may say indefensible; while the United States has first-class fortifications at Rouse's Point, on the shores of Lake Champlain, within a short distance of Canada's chief city, Montreal, capable of holding large forces.

No one has yet devised a plan by which the southern side of these three Provinces can be defended. Some time previous to their union Colonel Jarvis, an Imperial officer, made a report on the defences of Canada. He recommended the construction of some works in the Province of Quebec, and said: "unless these works are constructed it is worse than useless to continue any British force in Canada." The works recommended have not been constructed, and England has withdrawn her forces.

Since Colonel Jarvis made his report the United States has added greatly to her strength in the event of war, while Canada has done but little to strengthen her position

And the Canadian plains, known as the North-west territories, in which the Province of Manitoba and the district of

Keewatin lie, are, if possible, still more exposed to United States forces than the eastern Provinces. These territories are an extension northward of the great desert and prairie regions of the United States. In Canada they lie nearly in the form of a semicircle, having the international boundary line, the parallel of 49°, for a length of eight hundred miles, as their cord. On the east the Canadian plains are separated from Ottawa by a thousand miles of laurentian mountain region, and on the west a "sea of mountains" five hundred miles in breadth lie between them and the Pacific. In another part of this work we have more fully described these three sections of the Dominion, through which Canada purposes construct-ing a railroad.

Thus a door is open 800 miles wide, through which the armies of the Mississippi and Missouri valleys could freely pass to the Canadian settlements in the plains. There are no natural barriers between the prairie settlements in the States and those of Canada to prevent the ingress of United States troops; nor are there any rallying points on the plains where troops could take refuge. For a distance of 2,000 miles from the Atlantic westward the international boundary is close to the laurentian region. The chief part of the great lakes is in the United States. The latter owns Lake Michigan, two-thirds of Lake Superior, and half of Lakes Huron, Erie and Ontario, and their connecting rivers; and by the Washington treaty the free navigation of the River St. Lawrence to the ocean is ceded to that power forever. Hence the main line of railroad from Halifax westward, through the Provinces, is in many places close to the international boundary; and the Canadian Pacific railroad, if ever built, will be of but little benefit to the Dominion in the event of war. In the region of the great lakes it will in places be close to the frontier. In the Red River valley and great plains generally it will be peculiarly exposed to destruction. Indeed, the chief lines of Canadian railroad are liable to be destroyed in many places in a few days or even hours after a declaration of war.

And by the San Juan award the United States may prevent Canadian shipping entering the Strait of Georgia. This sub-ject has recently engaged the serious attention of the Cana-dian Government and the Chief Engineer of the Pacific rail-road. Application was made to the British naval officers who had been stationed at this part of the Pacific coast for information as to the character of the harbours in a nautical point of view, and also in regard to defence." From the naval testimony furnished, taken in conjunction with the admiralty charts, the following are some of the deductions drawn by Mr. Fleming in his report of 1877:—

"That as far as known, Burrard Inlet, an arm of the Strait of Georgia, is the best harbour and the easiest of approach from the ocean.

"That the Strait of Georgia is separated from the ocean by two archipelagos, one to the north, the other to the south of Vancouver Island.

"That the approach by the north of Vancouver Island to the Strait of Georgia is hazardous and objectionable.

"That the approach by the south of Vancouver Island is through passages more or less intricate, between or at no great distance from islands known as the San Juan group.

"That the most important islands of the San Juan group are in the territory of a foreign power, and that from their position they hold the power of assuming a threatening attitude towards passing commerce."

This group, known as San Juan, Stuart, Patos, Lopez, Fidalgo, and other islands, lie between Vancouver Island and the main land on the south. "All the naval authorities," says Mr. Fleming, "admit that vessels on their course to Burrard Inlet, Howe Sound, or Bute Inlet, would be exposed to the guns of the United States in the event of hostilities, and that the navigation of the channel would greatly depend on the force of the United States in the locality."

In a subsequent report, Rear Admiral A. De Horsey, who made a nautical survey of the approaches to the inner terminus of the Canadian Pacific railroad, says that the San Juan and Stuart islands "form the key of the navigation inside Vancouver Island. In case of war with the United States that power might readily stop our trade through Haro Strait." He says San Juan was visited in September, 1877, by General Sherman, of the United States, with a view to its fortification, and further remarks that "the possession of San Juan might enable that country in case of war to cut off our supply from the coal mines by sea." And the Victoria *Colonist* says that "a single battery erected on Lopez Island would command Rosaro Straits, and a speedy end would be put to Canadian commerce on the Pacific should hostilities between Great Britain and the United States ever occur."

23

There are some good harbors on the outer coast of Vancouver Island, but "the surveys have....clearly shown," says Mr. Fleming, "that the bridging from the main shore to Vancouver would be unprecedented in magnitude, and that its cost would be indeed enormous." Besides, the extension of the Pacific railway to the outer coast of Vancouver would involve the construction of 234 miles of additional railroad, the greater part of it passing through mountainous regions, besides the bridging of Seymour Narrows, which is all but impossible. And the selection of Burrard Inlet as the Pacific terminus of the Canadian Pacific railroad renders that end of this road peculiarly exposed to destruction in the event of war with the United States.

Thus the unfortunate decision of the international boundary so far north from beginning to end, together with the isolation of her chief settlements, will ever denationalize the Dominion of Canada. With such frontiers integrity of territory could not be maintained in any other part of the world.

We know that there is promulgated in the United States a policy of territorial extension of her frontiers. And remembering the hostile attitude she has assumed towards Great Britain and these Provinces in the past, and viewing the public questions between the States and Provinces still unsettled, and the ceaseless intercourse which necessity compels Canadians to hold with them, we have, it must be confessed, but little confidence in the permanence of peace. The position of these adjoining countries is not like that of two equally powerful nations whose fear of war might be mutual. Peace with a more powerful nation, according to the teachings of history, is secure only so long as the interests or the ambition of that nation permit it to last. Immediately after assuming her independence, the United States began, after the fashion of the nations of Europe, to extend her boundaries. At no period in history was there a stronger desire among the nations of Europe and America than at present to enlarge the home domains. Nearly all the feeble nations whose independence Britain guaranteed have become incorporated, as we have seen, with great powers. Such appears to be the manifest destiny of feeble communities throughout the world. Britain, France, the United States, Prussia, Russia, Austria and Italy, are each the unification of a number of other nations or communities, combining under the influence of the same general law,

The continental nations of Europe, who, by reason of their close geographical contiguity, are compelled to keep immense forces under training, and strong fortifications on the salient points of their frontiers. "The great questions of our time," said Count Bismarck, "are to be decided not by speeches and resolutions but by blood and iron." Such concentrations of power preclude the possibility of any one nation being able to defend a number of isolated colonies, or even prevent the absorption of feeble nations by great powers. It is therefore no particular disgrace to any power, however great, not to be able to control the destinies of colonies and nations as formerly.

Never did such vast preparations for war exist as at present. Every nation in Europe maintains its forces on a war scale even in time of peace. And the form and power of war are very different from what they were in the days of Wolfe and Montcalm, or even from what they were half a century ago. Large concentrated populations, with powerful armies and navies, strong fortifications, aided by all the advantages secured by modern inventions, long experience, and immense revenues, make war terrible indeed, and highly destructive to feeble powers. Though the British army is comparatively small, her navy is powerful.

It is to be regretted that even the most civilized nations are not to be trusted when interest inclines them otherwise. In the recent war between France and Prussia the latter declared that she only made war against the Emperor, and not against France, but at the close of the war she made France pay an immense indemnity in gold, and retained two valuable Provinces as a forfeit. And even in the war between christian Britain and pagan China, in 1839, a leading British statesman said: "That justice was on the side of the pagan."

In a previous article we have shown that according to the enunciations of British and Canadian statesmen, the tie which binds Canada to the Empire is a mere nominal one, the Governor General being the only connecting link, and consequently Canada is virtually *independent*.

As a general rule the real independence of a country depends upon the exclusion of outside influence. Wherever the spirit of freedom exists it resents the intrusion of another power into its concerns at home as a violation of its freedom. How far Canadian independence comports with this principle is a question.

It would be well for Canadians to look at the question of defences fairly; it would be well to know whether Britain has given any guarantee that she will protect Canada; and in the next place it would be well to consider whether all the forces that Britain could spare and Canada could muster would be able to keep these Provinces from being overrun by the forces of the United States, and finally subjected to that power.

It is difficult to avoid the conclusion, considering the territorial and other valuable resources which Great Britain has ceded to the United States, and looking at the immense empire adjoining the Dominion which the Union has acquired from other nations, without an effort on the part of Britain to obtain even a part of these valuable territories, that the mother country did not set much value on her possessions in North America. And since the report of Capt. Palliser on the Northwest territories, that of Col. Jarvis on the defences of Canada proper, and Major Robinson's report of the Intercolonial railroad route, were laid before the Imperial Parliament, this view of the subject received a strong impulse. During the civil war in the United States the defenceless state of Canada was fully acknowledged. At the close of that war Great Britain urged these colonies to unite, and when united she withdrew her forces, and told the Dominion to defend itself. Britain's annual expenditure in defence of the Provinces, said Mr. Gladstone, "is a very heavy charge, and it is our duty in every way to get rid of it."

And Sir Charles Adderley, a distinguished member of the Conservative party in England, referring to the colonial system, said: "I believe the Canadians are much more likely to involve us in a war than we are to inflict one upon them;" and that Great Britain cannot undertake to defend the colonies "for the sake of the Canadians."

The Duke of Newcastle in a dispatch to Sir George Grey, Governor of New Zealand, said: "The cost of all war should be borne by those for whose benefit it is carried on."

And, referring to what is termed the glory Great Britain derives from an extensive colonial empire, Sir George Lewis remarked, "that a nation derives no true glory from any possession which produces no assignable advantage to itself or to other communities. If a country possesses a dependency from which it derives no public revenue, no military or naval strength, and no commercial advantages or facilities for

emigration which it would not equally enjoy though the
dependency were independent, such a possession cannot justly
be called glorious."

Adam Smith, in his work on "The Wealth of Nations,"
shows at great length that "if any of the Provinces of the
British Empire cannot be made to contribute towards the support of the whole Empire, it is surely time that Great Britain
should free herself from the expense of defending those Provinces in time of war and of supporting any part of their civil
and military establishments in time of peace."

The Right Hon. John Bright, formerly a member of the
British Cabinet, in his place in Parliament, asserted that
"there is no statesman in England who will venture to bring
about the shedding of one drop of blood" in defence of British
North America. And in his Rochdale speech he was still
more explicit on this subject. He said: "If a man had a great
heart within him, he might cherish the hope that, from the
point of land which is habitable nearest the Pole, to the
shores of the great Gulf, the whole of that vast continent
might become one great confederation of States, without a
great army and without a great navy; not mixing itself
up with the entanglements of European politics; without a
custom house inside through the whole length and breadth
of the territory, and with freedom everywhere; such a
confederation would afford at least some hope that man is
not forsaken of heaven, and that the future of our race may
be better than the past."

In the discussion of the British North America Act in the
British Parliament, Mr. Aytoun said: "He never had met
with any man not a member of the Government who considered that it was possible to defend Canada against an attack in
force by the United States."

Sir C. W. Dilke, in his work "Greater Britain," referring
to his visit to the chief places in Canada, says, p. 382: "As for
our so-called defence of the colonies, in war time we defend
ourselves; we defend the colonies only during peace. In war
time they are ever left to shift for themselves....The present
system weakens us and them, us by taxes and by the withdrawal of our men and ships." And, "were the Americans
as fully convinced as we ourselves are of our total inability
to carry on a land war with the United States on the western
side of the continent, the bolder spirits among them would
cease to feel themselves under an assumed necessity to show
us our own weakness and their own strength."

In 1871 Mr. Cardwell, in laying his scheme for defending
the British Islands before Parliament, said: "It is an almost
universally accepted principle of English policy, that in these
days it is no longer desirable to maintain at the expense of the
British tax-payer a standing army to defend our distant
colonies."

Sir Charles Adderley contends that "on English colonies
with representative government of their own there can be no
doubt about the mischief of intruding either home government or protection."

And this view of Britain's colonial policy has been reiterated
by her leading press. We only have space for a few citations.

When the decision by the Emperor of Germany in relation
to the Pacific end of the international boundary was published,
the London *Times* reminded Canada of its relations to the
Empire. "The San Juan difficulty is settled," said the *Times*.
"There are, however, certain considerations connected with
the question which merit attention. It must always be remembered that this is a Canadian matter."

Referring to the Washington treaty the Canadian Government, in a dispatch to the Secretary of the Colonies, said:
"The principal cause of difference between Canada and the
United States has not been removed,.... but remains a subject
for anxiety. That a cession of territorial rights of great value
has been made to the United States, not only without the previous assent of Canada, but contrary to the expressed wishes
of the Canadian Government." In the article above referred
to the *Times* says the British Commissioners "acted under the
direction and control of the Government at London, who were
consulted at every step of the negotiations....The men of the
Dominion say that they were sacrificed at every stage by
agents who were as ignorant as they were incompetent to
defend their interests. The British Columbia boundary is
gone. The claims for compensation for the Fenian raids—
claims which, according to the principles laid down by the
arbitrators at Geneva, could not be disputed—were given up on
the least show of resistance. Their fisheries have been sold
for a sum of money which remains to be fixed by a cumbrous
machinery very slow to be set in motion, and to be paid at
some distant day when the House of Representatives may vote
the sum which may be awarded. We know that the indigna-
tion of the inhabitants of Canada last summer was such that

we had to bribe the Legislature of the Dominion with a guarantee of £2,500,000 before its assent to the treaty of Washington could be procured."

The *Times* of course did not anticipate so early a settlement of the fishery claims as has since taken place; but the force of the great journal's remarks is hardly if at all weakened by the payment of the Commissioners' award.

It is true that the San Juan boundary, the fisheries, and the Fenian raids, are Canadian affairs, but it is equally true that Canadians had but a feeble voice in their settlement. Canadians might say to Britain, in the language of the *Times*: "You have abandoned our fisheries; you have sacrificed our frontier; you have not given us open trade with the States; you have not secured any satisfaction of our claims for wanton injuries." The writer concludes that all was done for Canada that could be done. That both Britain and the Dominion are "now in a false position, and the time has arrived when we should be relieved from it."

In another article, referring to the forces of the United States, the *Times* said that that country, containing "forty millions of people, rich, industrious, energetic and intrepid, must comprise resources which no enemy could venture to despise. The Americans had no army in 1860, but in 1862 they had hundreds of thousands in the field. They had no navy, and yet in a few months time they had produced a powerful fleet of new and formidable fighting ships. In the course of years they raised and spent upward of £600,000,000, and such is the magnitude of their national resources that what they did in the civil war they could certainly, on a similar impulse, do over again. . . . Their exposed points are comparatively few, their resources are unbounded, and, though they do not desire war, they would not be slow to accept a challenge."

It was this view of the resources of the States that led the *Times* to say: "We are quite aware that in the event of war we should not be able to render effectual aid to our Canadian Dominion, and that our fellow subjects out there would either have to fight at a terrible disadvantage, or mortify our pride by anticipating defeat and yielding to terms. In a national point of view that would be no loss to this country."

After a careful review of the utterances of Britain's statesmen and press in regard to British North America, we find it difficult to avoid the conclusion that the Confederation Act, the Washington Treaty, and the recent money guarantees, were the full development of Britain's policy in regard to these Provinces. She was thus enabled to cover an honorable retreat from the political concerns of this country. She found a fair excuse for withdrawing her "troops," which the London *Post* said "have been left too long in such dependencies already. The transitional condition of affairs is the only excuse for the delay that has occurred. The colonies have, as it were, been setting up house for themselves, and in the completion of their arrangements they have been permitted to avail themselves temporarily of the paternal resources."

However, in leaving the Dominion of Canada to itself, it must be obvious to those who have carefully watched the progress of events, that in the settlement of disputes, matters purely Canadian only occupied a secondary place in the councils of Great Britain. By the Washington Treaty Britain's own disputes with the States were fully settled, while "the principal cause of difference between Canada and the United States has not been removed,....but remains a subject for anxiety." At the close of Britain's political life in this part of the American continent her disputes with the States had assumed a threatening character, which created much uneasiness on both sides of the ocean. However, having territorial rights of great value adjoining the United States, which she had the legal right to cede, and which the latter desired to have, Great Britain was enabled to get all her disputes, except those affecting the Dominion, fully settled. Thus Great Britain was pleased to get the Alabama and other disputes so easily settled, and the United States was fully satisfied; but not so with Canadians, who declared that their interests had been sacrificed.

But few of the claims of Canada were maintained. Neither Britain nor the States would pay Canada's claims for Fenian raids. Hence the Canadian Government declined to recommend the Legislature to sanction the Treaty of Washington. At this stage Canadian weakness developed itself. On being told that the Imperial Government "would deeply deplore" such a course the Dominion accepted what the London *Times* was pleased to call a bribe; that is, England's name to a loan of two millions and a half sterling as the only compensation for the loss of vast resources. The old colonies revolted for

185

reasons of much less importance. A century ago an entirely different sentiment prevailed both in Britain and her colonies from that which now exists between her and the Dominion of Canada. Then the old colonies fought for their independence. Since that time England has been ceding to the same country large sections of what she had left, and urging the Canadians to take the residue and do what they please with it. By the Washington Treaty the manifest destiny doctrine evidently received a powerful impulse.

During the rebellion in the United States the public sentiment in Great Britain, France, and in these Provinces was chiefly in favour of the independence of the southern States; it was indeed in favor of the balance of power on this continent being more evenly adjusted. It was believed by many that the States would be permanently disunited; that they would be formed into several independent nations. "I cannot believe," said a leading English statesman, "that we shall see the same society and form of Government....as existed before the civil war."

Canadians would not have objected to a reconstruction of frontiers. They would not have objected to the doctrine propounded in the Canadian Legislature by Colonel Rankin, who said: "No force we can bring into the field, no line of forts we can construct, nor indeed any course that could be adopted would so effectually protect us, so absolutely guarantee the inviolability of our soil as the recognition of the southern States by Great Britain."

It is impossible, however, to foresee what might have been the full effect of such recognition upon the future of America. We might premise that if England and France had recognized the southern confederacy, the States would have been divided into two or more nations. France might have established a monarchy in Mexico; Canada might, in case the French Emperor did not reconquer Quebec, have been able to shape a national course, and thus keep the balance of power on this continent more evenly adjusted. In these events, however, Canada would have had a slave power of the worst stamp for an ally, and the northern Union as an inveterate enemy. The great ruler of nations has otherwise directed.

No two parts of the world are so capable of being free from warlike complications with each other as Europe and North America. Europe for the Europeans, and America for the

24

Americans, seems to be fair for both. Europe comprises a cluster of nations, each of which feels it necessary to be always ready for war. Any one of its great powers in the east may at any time, and on the most paltry pretext, disturb the peace of the whole. In North America, Mexico has enough and more than enough to do in governing her own people. If the United States and Canada would unite and become the head of a peace policy, an armed peace in North America would not be necessary. By such an honorable course both countries would avoid war and enormous war debts. But if history is to repeat itself, if the burdens of a military system are to be imposed upon the scattered settlements of the Dominion, if she has to fly to arms whenever the United States casts dark shadows, the future of Canada will be precarious.

The Canadian Minister of Militia in his report for 1875 forecasts the policy of the future. He says: "A small standing army this country, like all other countries desiring to hold a position in the family of nations, eventually in the nature of things will have." And the reason assigned for this warlike attitude of the Dominion is that "a nation without being backed by physical power would have but small influence, if any, in the politics and councils of the nations of the world."

It is difficult to see what a small standing army could do to make the voice of Canada heard and respected in the councils and politics of the United States. What Sir Brenton Haliburton said half a century ago is equally true to-day: "The present colonies in North America are differently situated from those formerly possessed by Great Britain, which now comprise the United States. When disputes arose between them and the parent state the popular leaders were animated by the prospect of erecting the country into an independent nation; but no reasonable man in these colonies can ever entertain any such view. We can never become sufficiently strong to stand alone, and must therefore continue our connection with Great Britain or form one with America." Since Sir Brenton wrote these words the Provinces have only added three millions to their population, while the States have increased more than thirty millions.

During the rebellion in the States the Government of the two Canadas took a similar view to that in the above citations. The English Government continued to impress " on Canada the necessity of making greater preparations as regards her defence." The Canadian Government replied: "That no probable combination of regular troops and militia would preserve our soil from invading armies; and no fortune which the most sanguine dare hope for would prevent our most flourishing districts from being the battle-field of the war." And "since the war of 1812 the United States have covered their country with a network of railroads, and that seven of these lines run directly in upon the Canadian frontier, while others traverse or reach the shores of the great lakes commanding the chief entrepots of Canadian commerce, and others again extend to the sea-board cities directly fronting the Province of Nova Scotia, or through the State of Maine, to within eighty miles of the borders of New Brunswick.... By the aid of these railroads it is obvious that the United States could at any time within a week concentrate upon their termini a hundred thousand men or more."

Since the date of these dispatches the population of the United States has increased in a ratio about fourteen times as great as that of the Dominion. And in no country are the means of transit more efficient than in the States. The extent of interior navigation has no parallel anywhere; the country is traversed by seventy-four thousand miles of railroad; telegraph lines are in all directions. Twenty lines of railroad connect with the Provinces, and others are being constructed.

All the chief settlements of the Dominion are contiguous to United States settlements. The former has no country out of which she can erect another tier of Provinces north of the present, and her chief settlements are disconnected by immense regions of uninhabitable country. Hence the facilities for sending United States troops into all parts of the Provinces are complete, while the geographical ties connecting the chief settlements in the Dominion might be severed in a few days after a declaration of war.

In the event of war the Dominion would have to depend upon Britain's army, navy and money. In return for British protection the Dominion might be willing to provide a regiment for service abroad, as on a former occasion. But as the cost, transport and armament of the regiment raised in Canada for service in the Crimea "was borne," says the Edinburgh *Review*, "by the mother country, it turned out in the end to be the most costly regiment in the service."

Britain's cost in sending troops to these colonies was very heavy. And the cost of transit and maintenance of a sufficient force to defend Canada against the United States would be enormous. The article of food alone would be a costly one. Great Britain depends for a large part of her food at home upon the Western States. Ontario is the only Province in British North America which produces more than sufficient food for its own inhabitants. Manitoba may at times produce more than enough for its own people. Consequently the Provinces are large purchasers of breadstuffs from the United States. Hence no war would entail more suffering, especially in these Provinces, than a war with the Republic.

The Dominion is the only country in America which offers to send troops to fight in Europe. In view of the United States policy of non-interference in the wars and political complications in Europe, a question might arise whether Canada's course in sending troops to fight in Europe might not give rise to complications in America.

With two countries situated like the United States and Canada, it is the easiest thing in the world to raise disputes, to revive international quarrels, especially while old grievances are not forgotten, while disputes remain unsettled, and new ones are continually arising. Indeed, it might not be difficult at almost any time to find a plea for war fully as strong as France had against Prussia in 1870, or Russia had against Turkey in 1877.

We may be told that Canadians will not do any act, or utter a word that will tend to disturb the present peace; that the United States is large enough already; that it is "decreed" that Canada shall be one of the nationalities in North America, and that if war come Great Britain will defend the Dominion. Those who take this view of the future of Canada do so regardless of the teachings of history.

Such is the state of Europe that no nation on that continent could fight against the United States without weakening its power and influence in Europe. America is a good country for European nations to trade with, but, as in the past, it is not a safe one for any of them to fight in. Great Britain is the most powerful nation in the world, yet if she had been at war with the United States in 1878 when Russia was settling her war account with Turkey, she might have been the weakest. Indeed, the signs of the times point to Europe and

not to America as the future theatre on which European
nations will do the most fighting. It is in India, and not in
Canada, where Britain's true interests lie. This conclusion
was foreshadowed by Great Britain's recent policy with regard
to Russia and the East. It is difficult to avoid the conclusion
that the recent money guarantees, the union of the Provinces,
the withdrawal of her troops and military stores, and her
ceasing to fortify these Provinces, fully show the chief end of
Great Britain's Canadian policy of the future. They show,
too, that the Dominion is independent and free to shape its
own destiny.

It is no part of Britain's policy to hold a country she intends
to protect in an utterly defenceless state until the actual occurrence of war, especially a country situated as Canada is with
regard to the United States. Great Britain is fully aware that
the Dominion cannot possibly defend itself. Then if she
intends to defend it why delay the erection of fortifications.
In her recent schemes for defence she seems to ignore all
responsibility for the defence of the Dominion. And the
Dominion itself has done comparatively nothing for its own
defence; consequently if defence be attempted, when war
breaks out it must be done in the presence of the enemy, with
endless perplexity and confusion.

From all the facts at command we might fairly premise that
Great Britain will not destroy her vast commerce with the
United States, she will not close her great flour market on this
side of the ocean by sending her forces to defend a defenceless
frontier in the heart of North America, especially when no
gain would result to her arms. In whatever light we view
the position and relations of the Dominion the future is not
reassuring. Britain has not the teeming millions and vast
resources here such as she has in the east to fight for. Her
fighting for Canada would be a forlorn hope indeed.

If Canada does not wish to follow the example of the feeble
nations of Europe whose close geographical contiguity to great
powers has caused the most of them to be incorporated with
the latter, it is absolutely necessary that she maintain friendly
relations with the United States instead of appealing to Britain
to protect her against that power. This is the only way to
secure the integrity of the Dominion of Canada. The question is, in the face of many conflicting interests, how is friend-
ship to be maintained?

As a step in the right direction, both countries might follow the example of the British isles, which have ceased boasting of victories over an alien people on the plains of Abraham and of Waterloo. The United States and Canada might cease boasting of victories over each other in the long past—cease wasting their resources in preparing to shed each other's blood in the future ; and thus pave the way for a united destiny—a future in repose and prosperity, without mingling in the complications and politics of other countries. That a union of these two countries on fair and equitable terms should be consummated is, no doubt, the desire of millions on this continent, and especially the prayer of hundreds of thousands in the States and Provinces who are united by the dearest ties of kindred.

It is possible that the present peace may continue beyond the province of all reasonable predictions. But peace may be environed with many of the attributes of war. If Canada shall be compelled to maintain an army, as her Minister of Militia recently said, if she has to build fortifications and maintain a navy, if a system of hostile tariffs is to be continued, and also two chief Legislatures, one at Washington and the other at Ottawa, each crippling the functions of the other, it is difficult to foresee how peace can be maintained. In order to exist as a nation Canada must expend the chief part of her revenue in preparations for defence, which, in view of her large and increasing debt, she cannot do. She must have powerful alliances which can be relied upon in time of need. In a word, if Canada cannot meet national responsibilities she may as well leave all preparations for war out of her calculations altogether; since in this age of steam, electricity, iron and money, the only safety is a constant and powerful preparedness for war.

In his recent report on the State of the militia of the Dominion, Lieutenant General Smythe says: "The broad experience of the world's history. . . . has never failed to prove that military protection is an indispensable though it may be costly insurance for the safety and independence of every nation." That "if any one thinks that good rule alone will restrain the hands of either foreign or domestic foes, or unaided make a nation respected in its sway, he has not advanced far in the study of human nature."

No doubt "public opinion" in the Dominion, as General

Smythe says, declares that "the militia vote is that most easily reduced." Indeed, appropriations for defence by the Parliament of Canada are generally viewed by the inhabitants of this country as money wasted, especially in the face of "hard times" and a large and increasing public debt. We act on the principle that our weakness is our greatest strength, and adhere to the vain delusion that nobody will interfere with us.

ERRATA.

Page 11, 2nd line from top, for "coast," read "east."
" 73, 3rd " " " for "east," read "west."
" 95, 2nd " " " for "78,636,756," read "75,728,641."
" 96, 18th " " " for "last year," read "in 1877."
" 96, 23rd " " " for "last two fiscal years," read "ending in 1877."
" 101, 33rd " " " for "1861," read "the latter year."
" 140, 30th " " " omit "St. Thomas....2,145."
" 144, for "78,000,000," read "7,870,000."
" 148, for "French," read "Froude."
" 155, 36th " " " for "states," read "slaves."
" 156, 35th " " " omit "two."

www.ingramcontent.com/pod-product-compliance
Lightning Source LLC
Chambersburg PA
CBHW030833270326
41928CB00007B/1037